T0189650

Studies in Big Data

Volume 5

Series editor

Janusz Kacprzyk, Polish Academy of Sciences, Warsaw, Poland
e-mail: kacprzyk@ibspan.waw.pl

For further volumes:
http://www.springer.com/series/11970

About this Series

The series "Studies in Big Data" (SBD) publishes new developments and advances in the various areas of Big Data-quickly and with a high quality. The intent is to cover the theory, research, development, and applications of Big Data, as embedded in the fields of engineering, computer science, physics, economics and life sciences. The books of the series refer to the analysis and understanding of large, complex, and/or distributed data sets generated from recent digital sources coming from sensors or other physical instruments as well as simulations, crowd sourcing, social networks or other internet transactions, such as emails or video click streams and other. The series contains monographs, lecture notes and edited volumes in Big Data spanning the areas of computational intelligence incl. Neural networks, evolutionary computation, soft computing, fuzzy systems, as well as artificial intelligence, data mining, modern statistics and Operations research, as well as self-organizing systems. Of particular value to both the contributors and the readership are the short publication timeframe and the world-wide distribution, which enable both wide and rapid dissemination of research output.

Nikos Karacapilidis

Editor

Mastering Data-Intensive Collaboration and Decision Making

Research and Practical Applications in the Dicode Project

 Springer

Editor
Nikos Karacapilidis
University of Patras and Computer
  Technology Institute & Press
  "Diophantus"
Rio Patras
Greece

ISSN 2197-6503   ISSN 2197-6511 (electronic)
ISBN 978-3-319-34996-1   ISBN 978-3-319-02612-1 (eBook)
DOI 10.1007/978-3-319-02612-1
Springer Cham Heidelberg New York Dordrecht London

Printed on acid-free paper

Springer is part of Springer Science+Business Media (www.springer.com)

Preface

Many collaboration and decision-making settings are nowadays associated with huge, ever-increasing amounts of multiple types of data, obtained from diverse sources, which often have a low signal-to-noise ratio for addressing the problem at hand. These data may also vary in terms of subjectivity and importance, ranging from individual opinions and estimations to broadly accepted practices and trustable measurements and scientific results. Additional problems start when we want to consider and exploit accumulated volumes of data, which may have been collected over a few weeks or months, and meaningfully analyze them toward making a decision. Admittedly, when things get complex, we need to identify, understand and exploit data patterns; we need to aggregate appropriate volumes of data from multiple sources, and then mine them for insights that would never emerge from manual inspection or analysis of any single data source. In these settings, "big data" analytics technology currently receives much criticism, in that it does not provide proper insight into what the data means. To make sense of big data and come with discoveries that help improve decision making in practical contexts, human intelligence should be also exploited. We need to provide the appropriate ways to nurture and capture this human intelligence in order to extract the necessary insights and improve the way machines deal with complex situations.

This book reports on cutting-edge research toward efficiently and effectively addressing the above issues. This research has been carried out in the context of an EU-funded FP7 project, namely Dicode (http://dicode-project.eu), which aimed at facilitating and augmenting collaboration and decision making in data-intensive and cognitively-complex settings. To do so, whenever appropriate, Dicode built on prominent high-performance computing paradigms and large data processing technologies to meaningfully search, analyze, and aggregate data existing in diverse, extremely large, and rapidly evolving sources. At the same time, particular emphasis was given to the deepening of our insights about the proper exploitation of big data, as well as to collaboration and sense-making support issues. Building on current advancements, the solution proposed by the Dicode project brings together the reasoning capabilities of both the machine and the humans. It can be viewed as an innovative "workbench" incorporating and orchestrating a set of interoperable services that reduce the data-intensiveness and complexity overload at critical decision points to a manageable level, thus permitting stakeholders to be more productive and effective in their work practices.

Chapter 1, "The Dicode Project" by Nikos Karacapilidis, introduces the overall context of the project, and reports on its scientific and technical objectives, the exploitation of its results and its potential impact. Moreover, it sketches the key success indicators of the project, together with the actions taken toward ensuring their accomplishment.

Chapter 2, "Data Intensiveness and Cognitive Complexity in Contemporary Collaboration and Decision Making Settings" by Spyros Christodoulou, Nikos Karacapilidis, Manolis Tzagarakis, Vania Dimitrova and Guillermo de la Calle, reviews the state of the art on collaboration and decision-making support in contemporary settings. Related issues concerning integration technologies are also discussed. The methodologies, tools, and approaches discussed in the chapter are considered with respect to the information overload and cognitive complexity dimensions. The chapter aims to provide useful insights concerning the exploitation and advancement of existing collaboration and decision-making support technologies.

Chapter 3, "Requirements for Big Data Analytics Supporting Decision Making: A Sensemaking Perspective" by Lydia Lau, Fan Yang-Turner and Nikos Karacapilidis, aims to advance our understanding on the synergy between human and machine intelligence in tackling big data analysis. It does so by exploring and using sense-making models to inform the development of a generic conceptual architecture as a means to frame the requirements of such an analysis and to position the role of both technology and human in this synergetic relationship. Two contrasting real-world use case studies were undertaken to test the applicability of the proposed approach.

Chapter 4, "Making Sense of Linked Data: A Semantic Exploration Approach" by Dhavalkumar Thakker, Vania Dimitrova, Lydia Lau, Fan Yang-Turner and Dimoklis Despotakis, presents an experimental study with a uni-focal semantic data browser over several datasets linked via domain ontologies, which is used to inform what intelligent features are needed in order to assist exploratory search through Linked Data. The chapter reports main problems experienced by users while conducting exploratory search tasks, based on which requirements for algorithmic support to address the observed issues are elicited. In addition, a semantic signposting approach for extending a semantic data browser is proposed as a way to address the derived requirements.

Chapter 5, "The Dicode Data Mining Services" by Natalja Friesen, Max Jakob, Jörg Kindermann, Doris Maassen, Axel Poigné, Stefan Rüping and Daniel Trabold, provides an overview of the data mining services developed in the context of the Dicode project. It addresses the usability of the services and indicates which big data technologies are being used to deal with very large data collections. It is shown that these services intend to help in clearly defined steps of the sense-making process, where human capacity is most limited and the impact of automatic solutions is most profound. This includes recommendation services to search and filter information, text mining services to search for new information und unknown relations in data, and subgroup discovery services to find and evaluate hypotheses on data.

Chapter 6, "The Dicode Collaboration and Decision Making Support Services" by Manolis Tzagarakis, Nikos Karacapilidis, Spyros Christodoulou, Fan Yang-Turner and Lydia Lau, presents a series of innovative services developed in the context of the Dicode project to facilitate and augment collaboration, sense-making, and decision making in knowledge intensive environments. The ultimate goals of the proposed solution are to make it easier for users to follow the evolution of an ongoing collaboration, comprehend it in its entirety, and meaningfully aggregate data in order to resolve the issue under consideration. A tool that enables the monitoring and investigation of the collective behavior of teams with respect to sense-making tasks is also presented.

Chapter 7, "Integrating Dicode Services: The Dicode Workbench" by Guillermo de la Calle, Eduardo Alonso-Martínez, Martha Rojas-Vera and Miguel García-Remesal, presents the innovative approach developed in the Dicode project regarding the integration of services and applications. A flexible, scalable, and customizable information and computation infrastructure to exploit the competences of stakeholders and information workers is presented in detail. The proposed approach pays much attention to usability and ease-of-use issues. The chapter reports on two major outcomes of the Dicode project regarding integration issues: the Dicode Workbench and the Dicode Integration Framework.

Chapter 8, "Clinico-Genomic Research Assimilator: A Dicode Use Case" by Georgia Tsiliki and Sophia Kossida, reports on the practical use of the Dicode platform in the biomedical research context. Through a real scenario, it is shown that the platform enables researchers to efficiently and effectively collaborate and make decisions by meaningfully assembling, mining and analyzing available large-scale volumes of complex multifaceted data residing in different sources. Evaluation results are included and thoroughly assessed.

Chapter 9, "Opinion Mining from Unstructured Web 2.0 Data: A Dicode Use Case" by Ralf Löffler, reports on the use of Dicode Workbench and Dicode services in the Social Media Monitoring context. Recognizing that Social Web has given the consumers a voice and Social Media has huge impact on brands and products today, the chapter discusses how the Dicode platform can support a collaborative work environment and offer technical solutions that improve the overall quality in the social media processes. Evaluation results are also included and assessed.

Chapter 10, "Data Mining in Data-Intensive and Cognitively-Complex Settings: Lessons Learned from the Dicode Project" by Natalja Friesen, Jörg Kindermann, Doris Maassen and Stefan Rüping, reports on practical lessons learned while developing the Dicode's data mining services and using them in data-intensive and cognitively complex settings. Various sources were taken into consideration to establish these lessons, including user feedbacks obtained from evaluation studies, discussion in teams, as well as observation of services' usage. The lessons are presented in a way that could aid people who engage in various phases of developing similar kind of systems.

Chapter 11, "Collaboration and Decision Making in Data-Intensive and Cognitively-Complex Settings: Lessons Learned from the Dicode Project" by Spyros

Christodoulou, Manolis Tzagarakis, Nikos Karacapilidis, Fan Yang-Turner, Lydia Lau and Vania Dimitrova, discusses practical lessons learned during the development of innovative collaboration and decision-making support services in the context of Dicode. These lessons concern: (i) the methodology followed for the development of the abovementioned Dicode services, (ii) the facilitation and enhancement of collaboration and decision making in data-intensive and/or cognitively complex settings, and (iii) related technological and integration issues. Detailed evaluation reports, interviews, and discussions within the development teams, as well as analysis of the use of the developed services by end users through the associated log files, provided valuable feedback for the formulation and compilation of these lessons.

The results of the Dicode project, as reported in this book, are expected to advance the state of the art in approaches on: (i) the proper exploitation of big data and the integrated consideration of data mining and sense-making issues, (ii) recommender systems, with respect to recommendations in heterogeneous, multifaceted data and the identification of hidden links in complex data types, (iii) understanding text to drastically reduce the annotation effort for extracting relations, (iv) Web 2.0 collaboration support tools in terms of interoperability with third-party tools and integration of appropriate reasoning services, and (v) decision-making support applications, by integrating knowledge management and decision making features as well as by building on the synergy of human and machine argumentation-based reasoning.

The advancements reported in the book shape innovative work methodologies for dealing with the problems of information overload and cognitive complexity in diverse collaboration and decision-making contexts. Both individual and collaborative sense-making is augmented through the meaningful exploitation of prominent data processing and data analysis technologies. The proposed solution is user-friendly and built on the synergy of human and machine intelligence. It masks the overall complexity of the underlying issues, thus allowing stakeholders to easily interact with large and complex data, providing them with meaningful recommendations upon which they can base their decisions and actions. Moreover, machine-tractable knowledge concerning the full life cycle of collaboration and decision making is accumulated and maintained.

Nikos Karacapilidis

Contents

Chapter 1
The Dicode Project

Nikos Karacapilidis

Abstract The Dicode project aimed at facilitating and augmenting collaboration and decision making in data-intensive and cognitively-complex settings. To do so, whenever appropriate, it built on prominent high-performance computing paradigms and proper data processing technologies to meaningfully search, analyze and aggregate data existing in diverse, extremely large, and rapidly evolving sources. At the same time, particular emphasis was given to the deepening of our insights about the proper exploitation of big data, as well as to collaboration and sense making support issues. This chapter reports on the overall context of the project, its scientific and technical objectives, the exploitation of its results and its potential impact.

Keywords Big Data · Collaboration · Decision Making · Sense Making · Data Mining · High-performance computing · Dicode FP7 project

1.1 Introduction

Individuals, communities and organizations are currently confronted with the rapidly growing problem of information overload [1]. An enormous amount of content already exists in the digital universe (i.e. information that is created, captured, or replicated in digital form), which is characterized by high rates of new information that is being distributed and demands attention. This enables us to have instant access to more information (that is of interest) than we can ever possibly consume. As pointed out in a recent IDC's White Paper [2], the amount of information created, captured, or replicated exceeded available storage for the first time in 2007, while the digital universe is expanding by a factor of 10 every 5 years.

N. Karacapilidis (✉)
University of Patras and Computer Technology Institute & Press "Diophantus",
26504 Rio Patras, Greece
e-mail: nikos@mech.upatras.gr

N. Karacapilidis (ed.), *Mastering Data-Intensive Collaboration and Decision Making*,
Studies in Big Data 5, DOI: 10.1007/978-3-319-02612-1_1,
© Springer International Publishing Switzerland 2014

People have to cope with such a diverse and exploding digital universe when working together; they need to efficiently and effectively collaborate and make decisions by appropriately assembling and analyzing enormous volumes of complex multi-faceted data residing in different sources [3–5]. For instance, imagine:

- A community of clinical researchers and bio-scientists, supported in their scientific collaboration by a system that allows them to easily examine and reuse heterogeneous clinico-genomic data and information sources for the production of new insightful conclusions or the formation of reliable biomedical knowledge, without having to worry about the method of locating and assembling these huge quantities of data (clinical and genomic data, molecular pathways, DNA sequence data, etc.).
- Or a community of clinicians, radiologists, radiographers, patients and pharmaresearchers being able to contribute more effectively to clinical decisions and drug testing by combining heterogeneous, collaboratively annotated datasets from patient results (e.g. blood tests, physical examinations, free text journals from patients on their experience from treatment) and different scan modalities (e.g. X-ray, Static and Dynamic MRI), without having to be anxious about tracking the data and their provenance through the complex decision making process, and the handling of the associated multimedia material.
- Or even, a marketing and consultancy company being able to effortlessly forage the Web (blogs, forums, wikis, etc.) for high-level knowledge, such as public opinions about its products and services; it is thus able to capture tractable, commercially vital information that can be used to quickly monitor public response to a new marketing launch; having the means to meaningfully filter, collate and analyse the associated findings; and use the information to inform new strategy.

The goal of the Dicode project (http://dicode-project.eu) was to turn this vision into reality. The project was funded by the European Commission under the FP7 Work Programme (contract number: FP7-ICT-257184). It started on September 1st, 2010 and its duration was 36 months. The partners of the Dicode consortium were: Computer Technology Institute and Press "Diophantus" (project coordinator, Greece), University of Leeds (United Kingdom), Fraunhofer-Gesellschaft zur Förderung der angewandten Forschung e.V. (Germany), Universidad Politécnica de Madrid (Spain), Neofonie Gmbh (Germany), Image Analysis Ltd (United Kingdom), Biomedical Research Foundation—Academy of Athens (Greece), and Publicis Frankfurt Zweigniederlassung der PWW GmbH (Germany).

This chapter describes the overall context of the Dicode project (Sect. 1.2), its scientific and technical objectives (Sect. 1.3), as well as the exploitation of its results and its potential impact (Sect. 1.4).

1.2 Overall Project Context

Many collaboration and decision making settings are nowadays associated with huge, ever-increasing amounts of multiple types of data, obtained from diverse sources, which often have a low signal-to-noise ratio for addressing the problem at hand. These data may also vary in terms of subjectivity and importance, ranging from individual opinions and estimations to broadly accepted practices and trustable measurements and scientific results. Their types can be of diverse level as far as human understanding and machine interpretation are concerned. At the same time, the associated data are in most cases interconnected, in a vague or explicit manner.

Additional problems start when we want to consider and exploit accumulated volumes of data, which may have been collected over a few weeks or months, and meaningfully analyze them towards making a decision. Admittedly, when things get complex, we need to identify, understand and exploit data patterns; we need to aggregate appropriate volumes of data from multiple sources, and then mine them for insights that would never emerge from manual inspection or analysis of any single data source. In other words, the pathologies of big data are primarily those of analysis. The way that data will be structured for query and analysis, as well as the way that tools will be designed to handle them efficiently are of great importance and certainly set a big research challenge.

In the settings under consideration, "big data" analytics technology currently receives much criticism, in that it does not provide proper insight into what the data means. To make sense of big data and come with discoveries that help improve decision making in practical contexts, human intelligence should be also exploited. We need to provide the appropriate ways to nurture and capture this human intelligence in order to extract the necessary insights and improve the way machines deal with complex situations.

Taking the above issues into account, the Dicode project aimed at facilitating and augmenting collaboration and decision making in data-intensive and cognitively-complex settings. To do so, whenever appropriate, it built on prominent high-performance computing paradigms and proper data processing technologies to meaningfully search, analyze and aggregate data existing in diverse, extremely large, and rapidly evolving sources. At the same time, particular emphasis was given to the deepening of our insights about the proper exploitation of big data, as well as to collaboration and sense making support issues. Building on current advancements, the solution proposed by the Dicode project brings together the reasoning capabilities of both the machine and the humans (Fig. 1.1). It can be viewed as an innovative "workbench" incorporating and orchestrating a set of interoperable services that reduce the data-intensiveness and complexity overload at critical decision points to a manageable level, thus permitting stakeholders to be more productive and effective in their work practices. Services that were developed and integrated in the context of the Dicode project are released under an open source license.

Fig. 1.1 The Dicode services exploit the cloud computing paradigm and build on the synergy of machine and human reasoning

The achievements of the Dicode project were validated through three use cases:

- *Clinico-Genomic Research Assimilator* The need to collaboratively explore, evaluate, disseminate and diffuse relative scientific findings and results is more than profound today. Towards this objective, Dicode elaborated an integrated clinico-genomic (tacit) knowledge discovery and decision making use case that targets the identification and validation of predictive clinico-genomic models and biomarkers (this use case is presented in detail in Chap. 8).
- *Trial of Clinical Treatment Effects* The goal of this case (which has been expanded in the second year of the project to cover broader clinical trials, not just for Rheumatoid Arthritis) was to facilitate the process of making clinical decisions in drug trials by combining datasets from patient results (blood tests, physical examinations) and the different scan modalities (X-ray, Static and Dynamic MRI scan images) to reveal the effectiveness of a drug within a trial.
- *Opinion Mining from Unstructured Web 2.0 Data* Through this case, we validated the Dicode services for the automatic analyses of the voluminous amount of unstructured information existing on the Web, especially in the highly dynamic social media space. Data for this case were primarily obtained from spidering the most popular social Web sites making use of APIs from various Web 2.0 platforms (this use case is presented in detail in Chap. 9).

1.3 Scientific and Technical Objectives

The project's objectives have been fully accomplished through an evolutionary approach characterised by:

- the active engagement of all stakeholders (technical partners and use case representatives) in the specification, design and evaluation of the foreseen technological solutions throughout the project;
- the adoption of an incremental development approach, which ensured that end users can experiment with the Dicode services from the early stages of the project (operational prototype versions of the Dicode services were available at the end of the first year of the project, enhanced versions were delivered in month 24, final versions were ready in month 33);
- the continuous refinement of user requirements through testing (involving users from all three use cases), and
- the early availability of an operational integrated suite of services, which facilitated trials and proof-of-concept purposes, enabled proper exploitation and dissemination activities, and ensured project sustainability.

The association between the project's objectives and the project's milestones is illustrated in Fig. 1.2. As shown, "Laying Foundations", "Integration, Validation and Enhancement" and "Getting Broader" was the overall goal for each year of the project, respectively. As justified in the following, the Dicode project successfully reached these goals.

In particular, the project's scientific and technical objectives were:

- O–1: *To fully understand the current practices and needs of diverse communities and organizations as far as data-intensive and cognitively-complex collaboration and decision making is concerned.* Three representative use cases were continuously elaborated throughout the project. Related settings were also considered, aiming to reveal practices and needs associated with both large data sets and real-time data (see Chap. 3). The accomplishment of this objective was critical for the applicability of the Dicode approach in a wide variety of settings.

This objective was of high importance throughout the project. Thoroughly considering the feedback from the two evaluation rounds of Dicode services across the project's use cases, an analysis of the lessons learned was documented and services' specifications were revised to inform the final iteration of development. A much deeper understanding of the use cases' differences and similarities, as well as of their potential to explore the full range of Dicode services, was achieved through close collaboration between technical partners and end users.

- O–2: *To provide a suite of innovative, adaptive and interoperable services (both at a conceptual and a technical level) that satisfies the full range of the associated requirements.* The development of Dicode services facilitated and augmented collaboration, sense-making and decision-making in data-intensive and

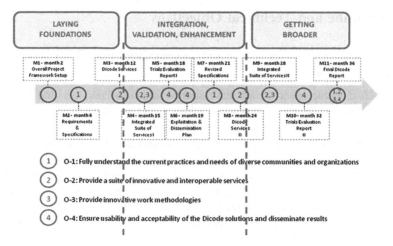

Fig. 1.2 S&T objectives, project's milestones and goals set for each year of the dicode project

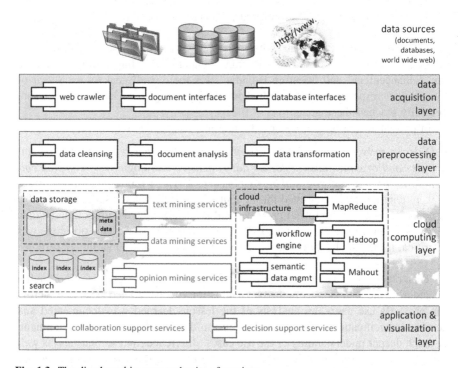

Fig. 1.3 The dicode architecture and suite of services

cognitively-complex settings, while also serving the underlying requirements of capturing, delivering and analyzing pertinent information (Fig. 1.3). Dicode services are running on the Web. Throughout the project, much attention was given to the adaptability of Dicode services with respect to changes in user requirements and operating conditions. Moreover, especially during the third year of the project, development efforts paid much attention to usability issues. Particular sub-objectives concern the development and seamless integration of:

- O–2.1: *Data acquisition services,* which enable the purposeful capturing of tractable information that exists in diverse data sources and formats. Particular attention was paid to web resources and the integration of social media APIs and high quality third-party feeds.
- O–2.2: *Data pre-processing services,* which efficiently manipulate raw data before their storage to the foreseen solution. Transformation of different kinds of documents into a canonical form, structuring of documents from layout information (e.g. detection of navigation, comments, abstracts), data cleansing (e.g. removing noise from web pages, discarding useless database records), as well as language detection and linguistic annotations are some of the functionalities falling in this category of services.
- O–2.3: *Data mining services,* which in many cases exploit and are built on top of a cloud infrastructure and other most prominent large data processing technologies to offer functionalities such as high performance full text search, data indexing, classification and clustering, directed data filtering and fusion, and meaningful data aggregation. Advanced text mining techniques, such as named entity recognition, relation extraction and opinion mining, help to extract valuable semantic information from unstructured texts. Intelligent data mining techniques elaborated include local pattern mining and similarity learning.
- O–2.4: *Collaboration support services,* which facilitate the synchronous and asynchronous collaboration of stakeholders through adaptive workspaces, efficiently handle the representation and visualization of the outcomes of the data mining services (through alternative and dedicated data visualization schemas), and accommodate a series of actions for the appropriate handling of data in each use case.
- O–2.5: *Decision making support services,* which augment (both individual and group) sense-making and decision-making by supporting stakeholders in locating, retrieving and arguing about relevant information and knowledge, as well as by providing them with appropriate notifications and recommendations.

This objective was of paramount importance for the success of the project. Taking into account feedback from the two evaluation rounds of the project, as well as recommendations of the Project officer and Project Reviewers, the final operational versions of the Dicode Workbench and integrated Dicode services were developed and tested across the project's use cases. Much attention was given to the openness of the Dicode solution, in order to augment exploitation purposes. An appropriate infrastructure of in-house computer clusters for running large scale data mining experiments and testing prototype implementations, as well as data collections for benchmarking based on textual and structured data, were set and maintained. Standards and guidelines for the development of Dicode services—aiming at ensuring interoperability between the services to be developed and reusability of them through diverse scenarios of use—were defined and revised upon the evolution of the project. Issues around both the conceptual and technical integration of the full range of Dicode services were thoroughly elaborated to upgrade user experience. According to the workplan, the final versions of the Dicode Data Mining Services (see Chap. 5), the Dicode Collaboration Support Services (see Chap. 6), the Dicode Decision Making Support Services (see Chap. 6), and the Dicode Workbench (see Chap. 7) were produced. In addition, a set of practical lessons learned while developing the Dicode's services and using them in data-intensive and cognitively-complex settings were reported. These lessons concern experiences, concrete recommendations and best practices from the development of the project's services, and they have been presented in a way that could aid people who engage in various phases of developing similar kind of systems (see Chaps. 10 and 11).

- O–3: *To provide innovative work methodologies that exploit the abovementioned suite of services and advance the current practices in terms of efficiency, creativity, as well as time and cost effectiveness.* These methodologies take into account the nature and needs of contemporary organisations and communities operating in a knowledge-driven economy.

This objective was highly important throughout the project. The established consensus on the role of the envisioned suite of Dicode services was significantly augmented through the two rounds of validation of the Dicode services, which provided valuable insights for the shaping of novel methodologies to be followed in stakeholders' daily work practices. During these evaluation rounds, a long and diverse set of end users tested the Dicode solution (services and Workbench) and provided valuable feedback by pointing out both strengths and weaknesses. These were considered through various real-world scenarios, which actually constituted the base for the definition of Dicode's innovative work methodologies (see Chaps. 8 and 9). The proposed methodologies reflect our experiences gained from the overall validation of the project's results and provide useful suggestions and insights to relevant communities and organizations.

- O–4: *To ensure usability and acceptability of the above services and work methodologies through their validation in real use cases, and disseminate the project's results by dedicated actions.*

This objective was also of paramount importance for the success of the project. Two rounds of evaluation of the Dicode Workbench and integrated services through the project's use cases were performed. Properly formulated metrics and questionnaires were employed to analyse the feedback received. Appropriate video-casts—based on everyday user stories from user communities, developers and early adopters—were prepared for each use case. The parameters assessed for each service concerned their acceptability, ease of use, usability, and overall quality (see Chaps. 8 and 9).

In addition, a comprehensive exploitation and dissemination plan has been produced, ensuring the impact and sustainability of the Dicode outcomes. Initial dissemination and exploitation activities included the development of a corporate identity of the project, the set-up of a web portal, and initial public relations efforts. A significant number of publications have resulted out of joint work among consortium members. These publications appear in international scientific journals and proceedings of international peer-reviewed scientific conferences and workshops (a detailed list of Dicode's dissemination activities appears at http://dicode-project.eu/index.php?q=news). Presentations of project-related work were also given in some of the top technology and marketing conferences. Moreover, Dicode organized four scientific workshops, one in the context of the world leading conference on collaboration support (CSCW 2012), another in the context of the best European conference on machine learning and knowledge discovery (ECML-PKDD 2012), a third one at the leading international conference on knowledge engineering and knowledge management (EKAW 2012), and a fourth one at world leading conference on hypertext and social media (Hypertext 2013). A series of exploitation activities has been also carried out, especially during the last 2 years of the project. Each Dicode partner put much effort in developing a concrete and realistic exploitation strategy (see Sect. 1.4). Several success stories concerning exploitation of Dicode results, development of strategic partnerships with industry and co-operation with other EU projects have been already reported.

1.4 Exploitation of Results and Potential Impact

The combination of academic and industrial partners within the Dicode consortium was perfectly suited for working with existing customers and collaborators in a variety of industry and academic segments to develop the Dicode platform for market use. Suitable targets were defined in the early stages of the project. As each target has specific needs which can be met through the technology developed in the Dicode project, partners in the project consortium were involved in cultivating and extending ties to their existing customer base to keep these key assets informed of project developments. The project partners also organized dedicated demonstrations of running prototypes and scenarios of use for key persons in the target organisations.

Figure 1.4 gives an overview of Dicode's target groups for the exploitation of the project's foreground. In the public sector, the focus lies on public services, public health and e-Science. In the private sector, advertising and communication, media and medicine are the main target areas. In the IT industry, Dicode caters to service integrators, service developers and consultancy companies.

Fig. 1.4 Dicode targets a wide audience

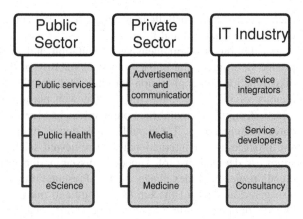

To ensure the sustainability of the Dicode project, each consortium partner formulated a detailed exploitation plan and carried out a set of associated activities, based on modern marketing and communication best practices. Market entry strategies followed in the context of the Dicode project included:

- The definition of appropriate targets, both public and private entities, and partners in the network of the consortium partners who have an interest in the outcomes of the Dicode project, and are also suitable for obtaining first experiences and willing to be used as success stories;
- Building out additional use-cases to fit the needs of the defined targets;
- Strategic partnerships with established players in the market.

The key success indicators of the Dicode project, together with their high effects and actions taken towards ensuring their accomplishment are summarized in Table 1.1 The final results of the Dicode project advance the state-of-the-art in approaches on (i) the proper exploitation of big data (dealing with the "big data fallacy" issue) and the integrated consideration of data mining and sense-making issues, (ii) recommender systems, with respect to recommendations in heterogeneous, multi-faceted data and the identification of hidden links in complex data types, (iii) understanding text to drastically reduce the annotation effort for extracting relations, (iv) opinion mining by considering opinion statements as n-ary relations and apply the highly scalable methodology implemented for their recognition (v) Web 2.0 collaboration support tools in terms of interoperability with third party tools and integration of appropriate reasoning and data mining services, and (vi) decision making support applications, by integrating knowledge management and decision making features as well as by building on the synergy of human and machine argumentation-based reasoning.

Such advancements have shaped innovative work methodologies for dealing with the problems of information overload and cognitive complexity in diverse collaboration and decision making contexts. Adopting the proposed solution, both individual and collaborative sense making are augmented through the meaningful

Table 1.1 Success indicators and actions to ensure them

Key success indicators	Actions
Deployment of the Dicode framework is not too costly → high rate of Dicode technology adoption	Adoption of standards, exploitation of existing prototypes and background technology, open-source policy
High level of the Dicode framework's acceptance by users involved in Dicode's use cases → increased users' productivity and creativity	Early and continuous involvement of end users in the development and evaluation of the Dicode platform
Acceptance of the Dicode framework by users outside the Dicode consortium → recognition of the Dicode platform's value from relevant groups and communities	Exploitation of Dicode partners' liaisons with scientific and business stakeholders; Dicode workshops and diverse dissemination activities; relevant market watch
Adaptability and proven portability of the Dicode framework in a wide range of application domains → acceptance of the Dicode framework by industry and academia	Generic and flexible development approach; show cases at relevant scientific and business events and stakeholders; Dicode scientific workshops; scientific dissemination activities; market analysis
Foundations for high-performance scalable data mining in the cloud computing initiative and related open source community → proven added value of the Dicode framework	Adoption of open source principles and concepts; advancement of cloud computing paradigm; dissemination activities related to dedicated workgroups and communities
Hit the optimal market → strengthened EU leadership in the domain of intelligent information management	Development of a detailed and coherent exploitation strategy and thorough consideration of associated perspectives

exploitation of prominent data processing and data analysis technologies. The Dicode solution is user-friendly and built on the synergy of human and machine intelligence. Adopting open standards, and in accordance with EU's recent initiatives on Open Systems and Data, the Dicode project has the potential of forming a rich ecology of domain specific and non-specific extensions. The Dicode platform allows for external data service providers to supply information, as well as for external developers to supply additional modules and applications, which are tailored to evolving market conditions. Finally, it enables diverse public and private entities to aggregate, structure, semantically enrich and analyse vast amounts of information. This turns the problem of information overload into a benefit of structured data, which can be used as the basis for decisions of better quality. Simply put, the Dicode solution is able to turn information growth into economic growth.

In particular, the potential impact of the Dicode project (including the socio-economic impact and the wider societal implications so far) concerns:

Better leveraging of human skills, improved quality and quantity of output and reduced time and cost allowing users to concentrate on more creative and innovative activities.

- The Dicode integrated suite of services and corresponding work methodologies facilitate and enhance the integration and aggregation of different stakeholders' perspectives across different collaboration and decision making activities, by explicitly addressing their knowledge and social dynamics. The Dicode platform is able to augment the creativity of stakeholders (stakeholders save time by skipping unnecessary tasks, accomplishing trivial tasks faster, while the platform provides a remedy to the information and cognitive overload). Stakeholders may easily customize the Dicode platform and concentrate on more creative and innovative activities.
- The Dicode platform enables new working practices for stakeholders involved in data-intensive and/or cognitively-complex settings. It has followed a component-based approach, based on open standards. This allows for further development by using and adapting existing modules, or developing new ones to cover the needs of related contexts.

Increased ability to identify and respond appropriately to evolving conditions (e.g. in finance, epidemiology, environmental crises…) faster and more effectively. Reinforced ability to collaboratively evolve large-scale, multi-dimensional models from the integration of independently developed datasets.

- In Dicode, machine-tractable knowledge concerning the full lifecycle of collaboration and decision making is accumulated and maintained. Consequently, the Dicode platform augments the productivity of stakeholders, e.g. by enabling them to easily locate and meaningfully reuse existing content. This affects both individuals and the workgroups they belong to.

- The Dicode platform improves the quality and quantity of the collaboration process. Since needs and user types evolve over time, the platform can be easily customized and adapted to address diverse needs and user types.
- The Dicode platform enhances collaboration between individual stakeholders through the meaningful integration and aggregation of independently developed applications (and associated datasets), which allows for a quicker consensus in the decision making process.
- The Dicode platform allows for external data service providers to supply information, as well as for external developers to supply additional modules and applications, which are tailored to evolving market conditions.

Higher levels of information portability and reuse by creating an ecology of systems and services that are dynamic, interoperable, trustworthy and accountable by design.

- The Dicode platform advances the state-of-the-art in information portability and reuse by considering interoperability issues, while also fostering standards-based integration and exploitation of information resources across organisational boundaries.
- The Dicode platform has been developed using existing standards and exploiting existing open source software.
- The Dicode project has developed a large-scale data processing platform. This platform allows for diverse data processing modules to be integrated through appropriate interfaces.
- The Dicode platform exploits, whenever appropriate, a cloud computing environment, which allows for improved information portability and reuse.
- The Dicode platform is web-based. This allows system independence for the end user.
- During the development of the Dicode platform, strong cooperation with committees and organizations which set standards in the fields of cloud computing was held. A series of contributions to free software projects concerning large-scale data processing in a cloud environment have been performed.
- Being designed with "openness" in mind, the Dicode platform is able to create a rich ecology of domain specific and non-specific extensions.

Increased EU competitiveness in the global knowledge economy by fostering standards-based integration and exploitation of information resources and services across domains and organisational boundaries.

- The global knowledge economy demands no barriers to entry. Accordingly, the Dicode platform has been developed by adopting open standards. Additionally, the platform allows for easy sharing of data and information. This enables the creation of marketplaces for information and information suppliers. Data rich applications can be implemented more quickly due to easy access of data through a shared environment.

- The web-based development of the Dicode platform, together with the exploitation of the cloud in some of its modules, allows for global access to innovative data processing services. Moreover, the platform can be easily adapted for international use (i.e. no cultural barriers to entry). The above may reduce fixed costs for companies using the Dicode platform, allowing them to invest more resources and money into their core line of business activities, thus providing them a competitive advantage in the international marketplace (i.e. no financial and technological barriers).
- The Dicode platform allows public and private entities to aggregate, structure and analyse vast amounts of information. This turns the problem of information overload into a benefit of structured data which can be used as the basis for better and quicker decisions. The Dicode platform helps stakeholders enrich current information, and turns the problem of information overload into knowledge discovery.

Strengthened EU leadership at every step of the computer-aided information and knowledge management lifecycle, creating the conditions for the rapid deployment of innovative products and applications based on high quality content.

- The European IT landscape is generally comprised of small and medium sized enterprises (SMEs). For many SMEs, it is difficult to develop new, data rich applications from scratch, basically due to the associated high investment costs. Open Source solutions, such as the Dicode platform, will reduce the barriers for SMEs in the development and hosting of data rich applications.
- In Europe, there are many different languages. For developing an application which can handle and process text sources from different EU countries, it is necessary to use different language dependent modules. Data and text processing standards, as supported by the Dicode platform, allow for the simple replacement of compatible modules which switch from one language to another (plug and play integration).
- Based on the existing Dicode infrastructure and services, new applications can be developed in less time. This yields to quicker "time to market" and faster return-on-investment due to decreased development costs.
- The Dicode platform is able to assist European companies in making better decisions quicker, based on the largest data set possible. As much of the data on the Web is text, Dicode solutions for issues such as sentiment analysis, opinion mining, data mining, trend mining etc. will continue to grow in importance for decision makers.

1.5 Conclusions

The Dicode platform enables a meaningful aggregation and analysis of big data in complex settings. The proposed solution (infrastructure and services), as described in detail in the following chapters of this book, allows for new working practices that turn the problem of information overload and cognitive complexity into the benefit of knowledge discovery. This is achieved through properly structured data that can be used as the basis for more informed decisions. Simply put, the Dicode approach is able to turn information growth into knowledge growth; it improves the quality of collaboration within a Web community, while enabling its users to be more productive and focus on creative activities.

As a last note, we point out that the overall Dicode approach is fully in line with a set of imperatives concerning challenges and opportunities with Big Data, which are reported in a recent White Paper authored by 21 prominent researchers [6]. Specifically, the Dicode platform enables stakeholders "to run heterogeneous workloads on a single infrastructure that is sufficiently flexible to handle all these workloads"; it is "designed explicitly to have a human in the loop", thus enabling "humans to easily detect patterns that computers algorithms have a hard time finding"; it provides "supplementary information that explains how each result was derived, and based upon precisely what inputs"; it offers "a rich palette of visualizations", which are "important in conveying to the users the results of the queries in a way that is best understood in the particular domain".

References

1. Eppler, M., Mengis, J.: The concept of information overload: a review of literature from organization science, accounting, marketing, MIS, and related disciplines. Inf Soc **20**(5), 325–344 (2004)
2. IDC. The diverse and exploding digital universe. White Paper. www.idc.com (2008). Accessed March 2008
3. Economist. A special report on managing information: data, data everywhere. Economist, London (2010)
4. Hara, N., Solomon, P., Kim, S.L., Sonnenwald, D.H.: An emerging view of scientific collaboration: scientists' perspectives on collaboration and factors that impact collaboration. J. Am. Soc. Inform. Sci. Technol. **54**, 952–965 (2003)
5. Shim, J.P., Warkentin, M., Courtney, J.F., Power, D.J., Sharda, R., Carlsson, C.: Past, present and future of decision support technology. Decis. Support Syst. **33**, 111–126 (2002)
6. Computing Community Consortium–Computing Research Association. Challenges and opportunities with big data: a community white paper developed by leading researchers across the United States. White Paper. http://www.cra.org/ccc/files/docs/init/bigdatawhitepaper.pdf (2012). Accessed Feb 2012

Chapter 2
Data Intensiveness and Cognitive Complexity in Contemporary Collaboration and Decision Making Settings

Spyros Christodoulou, Nikos Karacapilidis, Manolis Tzagarakis, Vania Dimitrova and Guillermo de la Calle

Abstract This chapter reviews the state-of-the-art on collaboration and decision making support in contemporary settings. Related issues concerning integration technologies are also discussed. The methodologies, tools and approaches discussed in the chapter are considered with respect to the information overload and cognitive complexity dimensions. The chapter aims to provide useful insights concerning the exploitation and advancement of existing collaboration and decision making support technologies.

Keywords Collaboration support · Decision making · Integration · Data intensiveness · Cognitive complexity · State-of-the-art

S. Christodoulou (✉) · N. Karacapilidis · M. Tzagarakis
University of Patras and Computer Technology Institute & Press "Diophantus",
26504 Rio Patras, Greece
e-mail: shristod@cti.gr

N. Karacapilidis
e-mail: nikos@mech.upatras.gr

M. Tzagarakis
e-mail: tzagara@upatras.gr

V. Dimitrova
School of Computing, University of Leeds, Leeds LS2 9JT, UK
e-mail: V.G.Dimitrova@leeds.ac.uk

G. de la Calle
School of Computer Science, Universidad Politécnica de Madrid, Madrid, Spain
e-mail: gcalle@infomed.dia.fi.upm.es

N. Karacapilidis (ed.), *Mastering Data-Intensive Collaboration and Decision Making*,
Studies in Big Data 5, DOI: 10.1007/978-3-319-02612-1_2,
© Springer International Publishing Switzerland 2014

2.1 Introduction

Information overload has become a major problem for today's organizations. While incoming data is rapidly increasing, making sense what is important for the current situation becomes difficult and time consuming. This becomes an even bigger problem in contexts where collaboration and decision making is taking place. These contexts are often associated with huge, ever-increasing amount of multiple types of data, obtained from diverse and distributed sources. In many cases, the raw information is so overwhelming that stakeholders are often at a loss to know even where to begin to make sense of it. Moreover, this data may vary in terms of subjectivity and importance, ranging from individual opinions and estimations to broadly accepted practices and indisputable measurements and scientific results. As broadly admitted, Big Data can negatively affect the effectiveness of decision making in an organization and create stress and cognitive overload to its stakeholders [1–3].

Being able to pull only the information relevant to a current problem solving scenario and efficiently share, interpret and use these information for decision making, becomes a challenge when the right tools and information systems are missing [4, 5]. When things get complex, we need to meaningfully aggregate large volumes of data from multiple sources. In many cases, the problem is not how to bring this information in the organization, but how to retrieve the information needed for a task from the information the organization already possesses [2].

Taking into account the above issues, this chapter reviews the state-of-the-art on collaboration and decision making support in contemporary settings. The remainder of the chapter is structured as follows: Sect. 2.2 outlines the overall decision making context, pointing to the need of synergy between human and machine reasoning. Section 2.3 reports in detail on state-of-the-art issues concerning collaboration and decision making support; strengths and weaknesses of the related technologies are highlighted, while recommendations related to their potential in addressing the information overload and cognitive complexity issues are also formulated. Section 2.4 is devoted to the state-of-the-art on integration technologies. Finally, Sect. 2.5 concludes the chapter.

2.2 Decision Making Support: On the Need of Synergy Between Human and Machine Reasoning

Decision Support Systems (DSS) first appeared in the late 1960s. They were defined as "interactive computer-based systems, which help decision makers utilize data and models to solve unstructured problems" [6]. Generally speaking, DSS research has focused on how information technology can improve the efficiency and effectiveness of a decision maker [7].

2.2.1 On the Evolution of Decision Making Support Technologies

In the 1970s and early 1980s, decision making support technologies were customarily focused on model development and problem analysis, while over the last two decades the related research has evolved to include additional concepts and views [8, 9] such as Group Decision Support Systems (GDSS) [10, 11], Executive Information Systems (EIS) and Knowledge-Based DSS. The advent of Internet/ Web and modern communication technology has resulted to the broadening of the organizational environment. Courtney [12] suggested that DSS researchers have to embrace a much more comprehensive view of the organizational decision making context and accordingly develop systems that are able to handle "softer" information. What started to evolve in the last few years is that issues related to the mental models of decision makers, expressing their organizational, personal and technical perspectives of the problem under consideration, are critical and have to be carefully addressed.

It is clear that the introduction of DSS received great attention from the beginning, since these systems were heading to important developments such as the integration of interactive systems for managers and professionals, the achievement of user-friendly environments, and the provision of a suitable framework for the handling of semi-structured and unstructured tasks. However, research on this area, having over-dealt with technological and definition issues (e.g., the differences between a DSS and an Expert System or an Executive Information System), has de-emphasized other major issues in improving decision making [13]. These issues include work structuring in order to improve coordination, use of communication technology to make decision making more efficient and effective, enforcing of rules and procedures for achieving consistency, and (semi)automation of data processing in data intensive decision making situations. Angehrn and Jelassi [14] have urged the DSS community to further consider the conceptual, methodological and application-oriented aspects of the problem. Conceptual focus is associated with the consideration of the nature of individual and organizational decision making processes, methodological focus with the integration of existing computer-based tools, techniques and systems into the human decision making context, and application-oriented focus with the consideration of the real organizational needs. Considering the above aspects, a series of prominent technologies has been proposed and evolved.

2.2.2 Prominent Decision Making Support Technologies

Data warehouses, on-line analytical processing, data mining and web-based DSS have been broadly recognized as technologies playing a prominent role in the development of current and future DSS [9, 15]. Data warehouses are subject-oriented, integrated, time-variant, non-volatile collections of data [16]; although they provide the infrastructure that enables businesses to extract, cleanse, and store vast amounts of corporate data [17], they do not provide adequate support for knowledge intensive queries in the organization. Data stored in a data warehouse are usually analyzed with the aid of on-line analytical processing (OLAP) tools [18]. Two basic types of OLAP tools are distinguished, namely Multidimensional OLAP (MOLAP) and Relational OLAP (ROLAP), each having its own advantages and disadvantages. A third type, namely Hybrid OLAP (HOLAP), attempts to combine the advantages of the first two. At the same time, web-based DSS deliver information and/or tools to a decision maker through a Web browser that is accessing the Internet or a corporate intranet [9].

The power of the above applications in processing vast amounts of data can be significantly augmented by data mining applications, built on concepts and techniques from AI and Statistics (such as Case- and Rule-Based Reasoning, Data Visualization, Fuzzy Analysis, and Neural Networks [19, 20]).

2.2.3 Paving the Way to the Dicode Approach

The above technologies certainly facilitate diverse aspects of decision making. Although there exist certain limitations in their suitability [21], they may aid DSS users to make better and faster decisions. However, there is still room for further developing the conceptual, methodological and application-oriented aspects of the problem. One critical point that is still missing is a holistic perspective on the issue of decision making. This originates out of the growing need to develop applications by following a more human-centric (not problem-centric) view, in order to appropriately address the requirements of the contemporary, knowledge-intensive organization's employees.

Such requirements stem from the fact that decision making has also to be considered as a social process that principally involves human interaction [22]. The structuring and management of this interaction requires the appropriate technological support and has to be explicitly embedded in the system. The above requirements, together with the ones imposed by the way decision makers work and collaborate today, delineate a set of challenges for further decision support technology development.

More specifically, as argued in [23], "a firm's only advantage in today's business environment is its ability to leverage and utilize its knowledge". Such knowledge resides in an evolving set of assets including the employees, structure, culture and processes of the organization. Of these, employee knowledge, and particularly tacit knowledge is identified as the dominant one, which is decisive at all mental levels and has to be fully exploited [24]. Such exploitation refers to the transformation of tacit knowledge to codified information, which is considered as a core process for economic activity and development [25]. The above advocate the adoption of a knowledge-based decision-making view [26], according to which, decisions should be considered as pieces of descriptive or procedural knowledge referring to an action commitment. On the other hand, in a decision making context the knowledge base of facts and routines alters, since it has to reflect the ever-changing external environment and internal structures of the organization [27].

The above mentioned integrated consideration of decision making and knowledge management can be further strengthened by the incorporation of features enabling decision makers to perform argumentation and experimentations on the issues raised. Many collaborative decision making problems have to be solved through dialoguing and argumentation among a group of people [28, 29]. In such contexts, conflicts of interest are unavoidable and support for achieving consensus and compromise is required. Each decision maker may have arguments in favor or against alternative solutions, as well as preferences and constraints imposed on them. Independently of the model used for decision making, argumentation is valuable in shaping a common understanding of the problem. It can provide the means to decide which parts of the information brought up by the decision makers will finally be the input to the model used. Moreover, argumentation may stimulate the participation of decision makers and encourage constructive criticism. To address the above category of requirements, a user-friendly discourse-based decision support environment should be developed.

The above discussion paves the way to the intelligent information management elaborated in the context of the Dicode project, which exploits and builds on the synergy of human (collective) and machine (artificial) intelligence, by giving them equal importance. Approaches primarily concerning the human intelligence, aiming to facilitate and enhance collaboration and decision making support in diverse settings, are considered in the next section.

2.3 Collaboration and Decision Making Support

2.3.1 Introduction

Collaboration and decision making support technologies play an important role in Dicode. More specifically, these technologies aim to: (i) offer a collaborative environment that allows users "immerse" in Web 2.0 interaction paradigms and exploit its enormous potential to collaborate through reviewing, commenting on and extending the shared content; (ii) maintain chains of views and opinions, accompanied by the supporting data, which may reflect, at any time, the current collective knowledge on the issue under consideration, and justify a particular decision made or action taken; and (iii) achieve group sense-making.

Many different categories of Web 2.0 collaboration and decision making support tools exist. Classified upon their basic purpose, the most popular categories of them are:

- Mind mapping tools;
- File sharing tools;
- Collaborative editing tools;
- Social networking tools;
- Note taking tools;
- Project and task management tools, and
- Argumentative collaboration tools.

In the following, we first outline a review of the above categories in an attempt to identify state-of-the-art functionalities and solutions offered by representative tools. We are particularly focusing on the cognitive overload issues these tools are prone to as well as the respective countermeasures/functionalities introduced to overcome potential cognitive overload situations. Then, adopting a community perspective, we sketch a review of community modeling, monitoring and adaptive collaboration support approaches.

2.3.2 Collaboration and Decision Making Support Tools

2.3.2.1 Mind Mapping Tools

These tools enable the creation and editing of "mind maps". A mind map is a diagram used to represent words, ideas, tasks, or other items linked to and arranged around a central key word or idea. Mind maps are mainly used to generate, visualize, structure, and classify ideas, and act as an aid to studying and organizing information, solving problems, facilitate sense-making, making decisions, and writing (http://en.wikipedia.org/wiki/Mind_map). A mind map could be seen as a depiction/overview of a certain piece of knowledge. It is a collection of "topics", formed around a central idea (could be a single word or a whole phrase) which forms the central topic. On the central idea, in a radial way, associated ideas/concepts are added. More formally, a mind map includes: (i) the central topic/idea which is unique for each mind map; it is the point where the map starts. (ii) main topics stemming from the central idea; each main topic is connected to the central idea through an associate line. (iii) main topics may be further analyzed to sub-topics; subtopics are of lesser importance. The central topic, main topics and subtopics form a connected graph.

Representative tools of this category are *MindMeister, Mindomo, Bubbl.us* and *XMind.* MindMeister (http://www.mindmeister.com) is an online tool basically used for brain storming. To enhance collaboration in data intensive environments, it supports a number of features, including: notifications via emails or SMS (to make users aware of changes), a focus (zoom in/out) feature (to browse and work on large maps), an expanding/collapsing tool for the subtopics of a topic, a filtering feature (to isolate part of the map fulfilling specific criteria), and a history tool (to retrieve different versions of the map).

Mindomo (http://www.mindomo.com) maps are in the form of a tree. To enhance collaboration in data intensive situations, Mindomo supports collapsing/ expanding of user selected parts of the map. Tagging is also possible by using a number of specific icons ("tags"), while a zoom in/out tool is offered to enhance browsing on large maps. The history tool provides a list of all the actions performed. The filtering feature may be used to isolate a part of the map fulfilling user-defined criteria. A search tool allows spotting of a topic or subtopic whose content matches the specified text.

Bubbl.us (http://www.bubbl.us) is designed to enhance brainstorming. To cope with information overload, Bubbl.us provides a zoom in/out tool to help browsing and scrolling in maps that include many bubbles. Collaborative editing of maps is also possible, resulting in sharing maps among the members of a group of users.

Finally, XMind (http://www.xmind.net) follows the classic format of a mind map. To deal with data intensive maps, XMind supports a filtering mechanism (selection of markers or labels). The extending/collapsing feature of the subtopics of a topic may also be useful when dealing with large maps. Boundaries are used as a topic aggregation mechanism.

2.3.2.2 File Sharing and Collaborative Editing Tools

File sharing tools refer to online tools which, at least provide a repository where each user may upload his files. Usually, uploaded files may be shared with other users, regularly backed up and restored at the user's will. Many file sharing platforms integrate online office suits that may be used to create, edit or view some types of the uploaded documents. Based on their emphasis on sharing documents, they are also used as collaborative editing environments.

Collaborative editing tools permit the joint authoring of documents via individual contributions. Wikis [30] are the most representative systems in this category. Wikis allow users to freely create and edit a Web page's content using any Web browser. They support a simple markup language with which new pages and cross-links between internal pages can be created. They have been used to support a great variety of tasks which include creating collaborative workspaces, managing shared knowledge and personal note taking. Wikis have been used as a collaboration platform to jointly author documents that reflect a group's understanding with respect to the issue under consideration.

Representative tools of these two categories are: *DropBox, Humyo, Box.net, Google Docs, MediaWiki, Confluence* and *PBworks*. DropBox (http://www. dropbox.com) is a web-based service enabling users to store and share files and folders through the DropBox server. Sharing of files with other users is based on an invitation model. To deal with information overload issues, especially in cases where many users collaborate and co-edit files, DropBox provides version history and file recovery. An online list provides access to all events that have taken place (such as file editing, time editing took place, name of the editor) in a shared folder.

Notifications inform users about changes on their web folders (e.g. when a shared file is changed by a user having access to the file).

Humyo.com (http://www.humyo.com) is an online storage service appropriate for sharing and synchronizing files across different computers. Humyo provides online "teamspaces", where files can be shared within teams. A filtering mechanism allows retrieval of files according to their content type (such as images, audio and video files). Searching of files is also available by specifying a number of criteria.

Box.net (http://www.box.net) is a cloud-based content management system. Its core services emphasize the sharing of folders and files via URLs. Sharing can be achieved by simply providing the email of users with which files and folders should be shared. In addition, Box.net offers features designed to enhance collaboration in data intensive environments, such as document versioning, file and folder tagging and filtering mechanisms.

Google Docs (http://docs.google.com) provides a number of features that may be useful in data intensive environments including history revision, searching, sorting based on file type and tagging. In addition, Google Docs supports user collaboration (sharing of documents or collaborative editing/creating of a document).

MediaWiki (http://www.mediawiki.org/wiki/MediaWiki) is a web-based software system that enables the collaborative authoring of web pages. It is one of the most widespread wiki software. To address data intensive issues and information overload, it provides features such as "watchlists", which enable the notification on changes to pages the user is interested in, page history and versioning.

Confluence (http://www.atlassian.com/software/confluence) enables the creation of "spaces" that permit grouping of content items. Confluence "pages" can be created and organized in hierarchies offering an effective categorization. Confluence includes a number of functionalities to minimize complexity in data intensive cases: notifications, separate spaces, access control mechanism, page categorization and tagging, as well as its searching and notification mechanism are useful features to avoid information overloading.

PBworks (http://pbworks.com), formerly known as PBWiki, is a wiki software emphasizing on business, education and personal usage. To cope with data intensive situations, it offers a number of features. Wiki versioning is provided, meaning that the administrator of the wiki is able to track all changes in the wiki's content since its creation. Notifications are sent via email whenever the wiki's content is changed. Each page of the wiki may be tagged with user selected tags. A keyword-based search mechanism is available for locating content/pages of the workspace.

2.3.2.3 Social Networking Tools

Social networking refers to the creation of social structures by connecting individuals with various types of ties such as friendship, kinship and common interests. Although a social network is possible to be established with personal contact, online social networking has become very popular with the development of social networking websites [31, 32]. Representative tools of this category include *Facebook, MySpace, LinkedIn* and *Twitter*.

Facebook (http://www.facebook.com) is an application to build large-scale social networks. To deal with the enormous number of its members and the vast amount of information contributed, Facebook incorporates the News Feed feature (a sum up of recent actions of Facebook members who are in some way related to a specific user). Notifications by email or SMS and creation of Facebook Groups (to "unite" a number of people sharing common interests) are also available.

MySpace (http://www.myspace.com) used to be the most popular social network until the advent of Facebook. To cope with information overloading, MySpace incorporates MySpace Groups. A space user has the option to get notifications. Searching for people/content is also possible through the provided search mechanism.

LinkedIn (http://www.linkedin.com) is as social network basically aiming at professionals and business market. Unlike other popular social networks, its main focus is to keep in contact colleagues, alumnus and in general people who share common professional interests. To deal with data intensiveness, LinkedIn has implemented the feature of LinkedIn groups. Each user is also able to create events and automatically propagate the event to all of his connections. A notification mechanism makes users aware of recent events in their networks (such as new connections).

Twitter (http://www.twitter.com) is a website that allows users to send and read short messages called "tweets". Its focus is on supporting "social grooming" or "peripheral awareness" i.e. making people aware of what the people around them are thinking, doing and feeling even when co-presence is not viable . To overcome the problems originating from the immense amount of messages exchanged between users, Twitter displays messages in reverse chronological order and enables organizing messages via "hashtags" (words in messages prefixed with the character #), which allow categorization of messages and function as links that display all messages belonging in that category.

2.3.2.4 Note Taking Tools

Note taking is about recording the most critical information out of a larger amount of information. The source of this information may be a lecture in a class, a project meeting or an everyday scheduling of a person's activities. Electronic note taking has to do with software tools that have been developed to enhance the process of note taking. Notes, apart from text content, may include files, multimedia content and worksheets. A user's notebook may be either personal or be shared with other online users through which collaboration can be achieved. Representative tools of this category include *Zoho Notebook, Evernote* and *SimpleNote*.

Zoho Notebook (http://notebook.zoho.com) supports the creation of notebooks that may contain various types of content. A Zoho Notebook user may own multiple notebooks, each one containing a number of different pages. To support collaboration in data intensive environments, Zoho NoteBook groups may be used to allow sharing content among specific users. Versioning of shared notebook content allows keeping track of changes and modifications.

Evernote (http://www.evernote.com) is an online tool designed to support note taking. A user is able to store permanently notes which may include text, images, audio files and handwritten "ink" notes. To cope with data intensive cases, EverNote Notes are organized in notebooks (being essentially "folders" of notes). Each Note may be tagged and organized in folders. A Search mechanism is also available for spotting a desired note.

SimpleNote (http://simplenoteapp.com) is a note taking application used for online storage of notes, lists and ideas. To enhance note management in accounts with a large number of notes, a searching mechanism has been developed. A note may have one or more tags that may be used either in searching or in listing. A note may be also pinned to rank notes based on their importance than the others and subsequently be moved on the top of the notes' listing.

2.3.2.5 Project and Task Management Tools

In general, project and task management software enables an integrated approach to manage planning, scheduling, monitoring, budgeting, resource allocation of large projects in an attempt to overcome the problem of using different software for each abovementioned process. Typically, these tools also permit content management, provide notifications via email and feature process related awareness mechanisms. Moreover, they provide services to enable cooperation and coordination of virtual teams. Representative tools of this category include *Basecamp*, *ActiveCollab* and *Redmine*.

Basecamp (http://basecamphq.com) is a Web-based project management system that focuses on making project management tasks easy to use. It provides milestone and deadline management, task-lists, wiki style content management, file sharing, time-tracking and messaging systems. With respect to data-intensiveness issues, Basecamp attempts to alleviate the effects of massive email exchanges which are frequent during project management tasks, by offering the "Message Board". The "Message Board" permits keeping project related messages and discussions centrally and accessible to all.

ActiveCollab (http://www.activecollab.com) is a Web-based project management system that provides milestone management, "tickets" (project tasks which are assigned to one or more project participants), checklists, time tracking, file and content management and per project discussion boards. ActiveCollab provides features such as awareness mechanisms (reminders about a ticket or a discussion), filters (on task assignments) and milestone "zoom-in" (on milestone details).

Redmine (http://www.redmine.org) is an open-source Web-based project management system that provides an issue management system, flexible role-based access control model, Gantt charts and calendars, time tracking, file sharing and content management system (Wiki), feeds and email notifications, as well as per project discussion forums. To manage the complexity of its projects, Redmine enables the creation of sub-projects. It also automates some tasks (e.g. Gantt charts and calendars are calculated based on the start and due time of issues). Awareness services are also available.

2.3.2.6 Argumentative Collaboration Tools

This category refers to software tools designed to help people take part in various types of dialogues in which arguments are exchanged. Such tools have been used in various domains such as commerce, education, law and planning [33]. Generally speaking, the design of software systems that can adequately address users' needs to express, share, interpret and reason about knowledge during an argumentative collaboration session has been a major research and development activity for more than 20 years. Technologies supporting argumentative collaboration usually provide the means for discussion structuring and visualization, sharing of documents, and user administration. They support argumentative collaboration at various levels. Furthermore, they aim at exploring argumentation as a means to establish a common ground between diverse stakeholders, to understand positions on issues, to surface assumptions and criteria, and to collectively construct consensus. Representative tools of this category include *Araucaria, DebateGraph, Compendium, CoPe_it!* and *Cohere.*

Araucaria (http://araucaria.computing.dundee.ac.uk) is a representative tool of this category, which enables argument analysis through diagrams. Araucaria arguments are being built by selecting phrases from a user defined text. Each selected text corresponds to a node on the argument diagram and lines (relationships) may connect one node (premise) to another (conclusion).

DebateGraph (http://debategraph.org) includes several mechanisms to support large scale argumentation and collaboration. For instance, there is a "history" mechanism for browsing the other users' actions, while a progressive visualization of the argumentation map, several awareness mechanisms and a search mechanism are supported.

Compendium (http://compendium.open.ac.uk) is a software tool designed for mapping information, ideas and arguments. Ideas on a Compendium map are expressed by using different types of nodes which are linked together with different types of relationships. To deal with data intensive environments, the tool includes a number of features such as the multiple-level maps (a map may include another map), the zoom in/out tool and the "aerial" view (to help in maps with a large number of nodes). A dedicated search mechanism has been also implemented. In order to reposition the nodes on a map, a user can multiple-drag nodes. A user is also able to store node "bookmarks" for easily locating a node on a large map.

CoPe_it! (http://copeit.cti.gr) is a Web 2.0 tool designed to enhance collaboration by sharing opinions and resources in communities of practice. A CoPe_it! user can create a personal or collaborative workspace, join and contribute to an existing workspace and add/share content through a workspace. Various argumentation items may be uploaded and linked. Users can collaborate in either an asynchronous or a synchronous way. CoPe_it! supports a number of features to enhance collaboration in data intensive cases. For instance, the "minimap" of a workspace provides an overview of its contents. Also, there is a "review/history" mechanism (enabling a user to follow the evolution of a workspace) and a filtering mechanism (based on multiple criteria).

Cohere (http://cohere.open.ac.uk) is an online visual tool used to create, connect and share ideas. The tool mimics the most popular social sites as it encapsulates the idea of people, people pages and groups. To deal with data intensive situations, it offers various filtering mechanisms. Tagging of ideas is also available, while a search mechanism may exploit the ideas' text, the associated tags, the name of a user or group and the text appearing on a connection.

2.3.3 Collaboration Tools Overview and Implications for Dicode

In order to devise a roadmap towards the identification of the appropriate collaboration technologies and functionalities to be further exploited in the context of Dicode, we categorize the tools discussed in the previous section according to several dimensions. In particular, we classify the categories and individual tools according to: (i) their collaboration objective (i.e. what the tools aim to achieve via collaboration?), (ii) functionalities that individual tools provide, (iii) the cognitive overload issues they are prone to, and (iv) the countermeasures that these tools introduce to overcome cognitive overload issues. These dimensions were important for the Dicode project, as they outline the solution space not only in terms of objectives and functionalities, but also in terms of technologies to remedy cognitive overload and data-intensiveness issues.

Table 2.1 classifies the presented categories of collaboration tools according to the objective they mainly aim for. We distinguish the following classes: (i) *community building*, when tools aim for locating people and building ties between them, thus forming communities, (ii) *communication*, when the objective is simply to enable peer-to-peer communication between participants, (iii) *coordination*, where the objective is to align the actions of a group of people to achieve a common goal, (iv) *sense making*, when tools aim at the exchange of opinions and ideas and generate meaning of the exchanged items, and (v) *decision making*, where the aim is to select a course of action among several alternatives.

Next, we classify each tool according to the services and functionalities it provides. Table 2.2 illustrates this classification (due to its simplicity and limited

Table 2.1 Collaboration tools categories and their objectives

	Community building	Communication	Coordination	Sense making	Decision making
Mind maps				x	
Collaborative editing			x	x	
Social networking	x	x			
Note taking				x	
Project/task management			x		
Argumentative collaboration		x		x	x

collaboration support, we do not include Twitter in the following discussion). The importance of this table is to show that although tools aim at specific objectives, this is not necessarily associated with specific services and functionalities, and that these tools provide a wide range of collaboration services which are not typical for their kind.

The set of services and functionalities that are examined has been selected based on their potential usefulness for the Dicode project and include: (i) *discussion*: the ability to facilitate exchange of ideas and opinions between groups of people (brainstorming) and make these ideas and opinions subject to comments, (ii) *archiving*: the ability to archive and organize the items under collaboration or the entire collaboration, (iii) *visualization*: the ability to provide advanced visualizations of the collaboration space such as graphs, (iv) *annotation*: the ability to annotate or tag resources in the collaboration space, (v) *chat*: the ability to enable peer-to-peer real-time interaction between participants, (vi) *awareness*: the ability to inform participants on the actions of other participants, (vii) *task lists*: the ability to maintain list of tasks along with computational support to evaluate and enforce these lists, (viii) *file sharing*: the ability to share files (documents and other resources) between participants of the collaboration, (ix) *document management*: to manage (organize) documents available in the collaboration space, and (x) *user and role management*: the ability to define users and roles through which the levels of access to the resources are controlled.

We then analyze collaboration tools categories according to the sources of cognitive overload (Table 2.3) and the countermeasures taken (Table 2.4). By the term 'source of cognitive overload' we refer to the characteristics of information that may lead to cognitive overload situations in each tool, while by 'countermeasures' the solutions that the tools make available to remedy cognitive overload. The analysis is based on the extensive list of causes and countermeasures reported in [34] (specifically, we elaborated those related to the Dicode project). More specifically, we identify the following sources of cognitive overload that need to be addressed by collaboration services in Dicode: (i) *rising number of information*: the information items brought into collaboration increase as the collaboration proceeds; such increase may not be gradual but may appear in bursts,

Table 2.2 Collaboration tools and their functionalities

	Discussion	Archiving	Visualization	Annotation	Chat	Awareness	Task lists	File sharing	Document management	User and role management
Mind maps		1, 2, 4	1, 2, 3, 4	2, 4	2	1, 2	2	1, 2, 3, 4	1, 2, 3, 4	1, 2, 3
Collaborative editing	9, 10				7, 8, 9, 10	5, 6, 7, 8, 9, 10, 11		5, 6, 7, 8, 9, 10, 11	5, 6, 7, 8, 9, 10, 11	6, 7, 8, 9, 10, 11
Social networking	12, 13			12, 13	12, 13	12, 13, 14		12, 13, 14	12, 13	12, 13
Note taking	15	16		16, 17	15	15		15, 16	15, 16, 17	15, 16
Project/task management	19					18, 19, 20	18, 19, 20	18, 19, 20	18, 19, 20	19, 20
Argumentative collaboration	22, 24	24	21, 22, 23, 24, 25	22, 23, 24, 25	22, 24	22, 23, 24		22, 24, 25	22, 24, 25	22, 24

Numbers correspond to tools according to the following mapping: *1* MindMeister, *2* MindDomo, *3* Bubbl.us, *4* XMind, *5* Drop box, *6* Humyo.com, *7* Box.net, *8* Google Docs, *9* MediaWiki, *10* Confluence, *11* PBWorks, *12* Facebook, *13* MySpace, *14* LinkedIn, *15* Zoho Notebook, *16* Evernote, *17* SimpleNote, *18* Basecamp, *19* ActiveCollab, *20* Redmine, *21* Araucaria, *22* DebateGraph, *23* Compendium, *24* CoPe_it!, *25* Cohere

Table 2.3 Causes of information overload for each category of collaboration tools

	Rising number of information items	Uncertainty of information	Info diversity and increasing no of alternatives	Ambiguity of information	Complexity of information	Intensity of information	Increasing dimensions of information	Information quality, value	Overabundance of irrelevant information
Mind maps	x	x		x	x		x		x
Collaborative editing	x	x		x		x	x	x	
Social networking	x					x			x
Note taking	x							x	
Project/task management	x				x				
Argumentative collaboration	x	x	x	x	x	x	x	x	x

Table 2.4 Countermeasures taken by collaboration tools

	Structuring information	Visualization (graphs)	Formalization	Simplicity	Customization and personalization	Levels of detail/ summaries	Awareness (notification, versioning)	Search and filtering
Mind maps	1, 2, 3, 4	1, 2, 3, 4	1, 2, 3, 4	1, 2, 3, 4		1, 2, 3		
Collaborative editing	5, 6, 7, 8, 9, 10, 11			5, 6, 7, 9, 11	9, 10	5	5, 7, 8, 9, 10, 11	6, 7, 8, 9, 10, 11
Social networking		12	12, 13, 14	12, 13, 14	12, 13, 14		12, 13, 14	12, 13, 14
Note taking	16, 17	15						16, 17
Project/task management	20		18, 19, 20			19	18, 19, 20	18, 19, 20
Argumentative collaboration	22, 23, 24, 25	21, 22, 23, 24, 25	21, 22, 23, 24, 25			22, 23, 24, 25	22, 24	21, 23, 24

Numbers correspond to tools according to the following mapping: *1* MindMeister, *2* MindDomo, *3* Bubbl.us, *4* XMind, *5* Drop box, *6* Humyo.com, *7* Box.net, *8* Google Docs, *9* MediaWiki, *10* Confluence, *11* PBWorks, *12* Facebook, *13* MySpace, *14* LinkedIn, *15* Zoho Notebook, *16* Evernote, *17* SimpleNote, *18* Basecamp, *19* ActiveCollab, *20* Redmine, *21* Araucaria, *22* DebateGraph, *23* Compendium, *24* CoPe_it!, *25* Cohere

(ii) *uncertainty of information*: the inability to assess quickly the relevance of the available information, (iii) *information diversity and increasing number of alternatives*: the situation in which diverse types of information exist and the number of solutions increases as the collaboration proceeds, (iv) *ambiguity of information*: the situation where information can be interpreted in several ways, (v) *complexity of information*: the degree of interrelationships of information, (vi) *intensity of information*: the importance of particular information items, (vii) *increase of information dimensions*: the situation in which the way the available information brought in during collaboration can be combined with an increasing number of other items or can be considered along different aspects and dimensions, (viii) *information quality and value*: the degree of worth of information and (ix) *overabundance of irrelevant information*: the excessive amount of irrelevant information which leads to a low signal/noise ratio of the items in the collaboration space.

From the above analysis, it results that a plethora of collaboration technologies is available, each of them aiming to support different objectives. Analyzing the tools with respect to the services they provide, it is evident that although tools belonging in the same category provide a common core set of services, they supplement them with services which are not typical for their category. While elaborating on the issue of sources of cognitive overload in each category of collaboration tools, it also revealed that all categories are prone to such concerns. However, the analysis also shows that each tool attempts to address the related data intensiveness and cognitive overload issues by introducing particular services or approaches, which aim at alleviating the severe consequences. In the same line, each category of tools favors particular cognitive overload countermeasures, which are explicitly designed to address the problems that occur in a particular collaboration context. When each tool is used independently, the available countermeasures may provide the required support to address information overload issues. However, when an integrated approach of the presented tools must be considered, i.e. when two or more tools have to be deployed to address collaboration needs, the countermeasures may be insufficient and of limited use. This is mainly because the countermeasures of each tool have a particular scope which is derived from the collaboration objective. Hence, tools which belong to different categories but exhibit common countermeasures conceive them in different terms, thus raising concerns on how to consider them when these tools have to be used jointly.

In Dicode, signals are strong that such an integrated approach to collaboration is required. In particular, in the project's context, argumentative collaboration, collaborative editing, note taking and mind mapping tools look promising to address the foreseen collaboration needs. Yet, these tools must be considered in an integrated manner. This integrated consideration of diverse collaboration tools raises questions on how to redefine the available countermeasures and adapt them in this new environment. In Dicode, we envisage collaboration tools grafted with effective cognitive overload countermeasures, which do not limit their focus to particular collaboration objectives but provide their services in situations where heterogeneous collaboration tools must interoperate.

2.3.4 Community Monitoring and Adaptive Collaboration Support

Virtual communities (VCs), where people with common interests and goal work together sharing experience, constructing collective knowledge, and taking collective decisions, are playing a vital role in the modern work practices in both business and academia. Stepping on this social phenomenon, Dicode intended to capitalize on the collective knowledge accumulated in virtual communities, as well as provide intelligent support to facilitate collaboration and decision making in such communities.

In a broad sense, VCs vary from loosely structured to closely-knit ones. An open, loosely structured community involves a large number of people with diverse interests, membership control is generally not imposed, and there are no restrictions of the interaction with the community information space. Examples of this kind of communities, such as forums or blog communities, are widely available on the web. Participation in such communities is on a voluntary basis. They are highly dynamic, include a broad range of participants, joining and disconnecting at any time. Research has shown that such communities are subjected to power-law distributions, and consist of overlapping clusters which can evolve over time [35, 36]. Loosely structured communities become informal drivers forming trends and broader influences.

In contrast, closely-knit communities involve a smaller number of people and usually exist in relatively well defined organizational or educational settings. This kind of communities are characterized by a common goal (e.g. planning experiments or making judgment on a patient's condition), shared interests among all members, some commitment to participate in collaboration activities, high level of interaction, and active participation, which sometimes involves pre-assigned roles. Closely-knit communities usually involve well-formed teams, e.g. people working on a research project or collaborative medical diagnostic teams assessing patients' conditions. These communities have controlled membership for accessing the community's space and resources, and are closed for the outside world. They have well-defined norms, responsibilities, and work in established trust and reputation models. Although tightly focused, closely-knit communities may be influenced by broader trends and developments in large, loosely coupled communities.

The community monitoring and adaptive community mechanism in Dicode concerns closely-knit communities. The monitoring mechanism focuses on the collaboration and decision making process within the closely-knit community, while the support mechanism can include support for internal interaction, collaboration, and decision making processes, as well as support from external, large, loosely structured communities.

2.3.4.1 Approaches for Community Modeling

Recent research trends look at intelligent ways to support the effective functioning of online communities. In this line, personalization and adaptation techniques play a crucial role. The effectiveness of personalized support provided to virtual communities depend on what is known about a particular community and in which areas the community may need support.

Modeling virtual communities has recently become very popular in different research areas. In user modeling, modeling group of members provides the grounds for generating group recommendations [37]. In social networks, community modeling aids the discovery of relationships between people and among communities [38]. We review approaches from both user modeling and social network below.

2.3.4.2 Discovering Connections

A fairly simple and elegant community model is presented in [39]. It is based on a list of topics based on the resources that VC members are sharing. A reward factor is calculated to measure the relevance of each contributed resource to the current topic the VC is working on. Each member has an individual user model consisting of the reputation measure of that member in the VC [39]. An earlier work in the same group presented a more elaborate relationship model [40], where users' interests are modeled based on how frequently and how recently users have searched for a specific area from the ACM taxonomy, and user relationships are derived based on any successful download or service that took place between two users. A more recent approach by Kleanthous and Dimitrova [41, 42] employs the metadata of the resources shared in the community along with an ontology representing the community context, and derives a semantically relevant list of interests for every user.

2.3.4.3 Modeling Interests

User interests have been extensively studied. For example, an approach where user interests are extracted as keywords from the user profiles and other web content shared by a user in the community is presented in [43]. An ontology is then accessed, where associations are derived with ontology concepts and further recommendations are made to users. Interests are also used in finding relationships between users or connections in social graphs. Other approaches extract interests based on tags users ascribe to items posted online [38, 44]. Relationships/associations between users are derived based on their tags. Members can then be connected by interest similarity between them. Another approach models user interests based on resources members are uploading or downloading [41, 42]. However, this exploits semantic enrichment of the uploading/downloading activities by using, in addition to the resource key words, concepts extracted from an ontology. This uses semantically-enriched data to extract interest similarity between community members.

2.3.4.4 Modeling Expertise

Interests of users are usually associated with expertise, especially in social network research [45–48]. The approach described in [48] extracts shared interests in a discussion based on posting/replying threads. Based on the discussion topics a member of the community is contributing to, his interests and expertise are extracted; subsequently, user interest relationships are obtained. A similar method, which is mining email communication networks, is followed in [46]. Relationships are inferred according to the expertise/interests of members, which are extracted from communication recorder in their email conversations. Modeling expertise relations plotted as graphs is also explored in [45]. A relational network is extracted according to people's publications. The expertise/interests of a person are obtained by his previous publications; and two people are considered related if they have publications in the same research area. Relevant to expertise is a person's influence in the community. This can be derived by applying social network formulas based on the community graph (e.g. see [42]).

2.3.4.5 Community Graph Models

Recent research employed graph theory to model communities and relationships between members [49, 50] or members' interactions in general [51, 52]. In [49], the individual user model represents the conceptual understanding of a user, based on which a graph network is constructed. Similarities are then extracted according to a user's conceptual understanding, and group models are derived based on the distance between members in a graph. The approach described in [50] uses the notion of interaction network to represent relationships between users in a learning community. Two members are related if they have modified the same resource; hence, they appear connected in the interaction graph. The approach described in [51] considers the exchange of messages as interaction between two users, represented in a graph. A relationship between two users exists if they have engaged in some message exchange [52]. Kleanthous and Dimitrova [42, 44] have developed community graph based on semantic relationships, in addition to the interactions between users; an edge connecting two members represents their semantic similarity to each other, and the relevance of this link to the community's domain.

2.3.4.6 Community Patterns

The community model can be analyzed to automatically detect problematic cases which can be used to decide when and how interventions to the community can be done, offering support to improve the knowledge sharing processes in the community [44]. Research in organizational psychology identifies processes that can have an impact on collaborative processes, and are important for the effective functioning of teams and closely-knit communities [53, 54]. Previous research has focused on three key processes: (i) *Transactive Memory*—members are aware how their knowledge relates to the knowledge of the others [55]; (ii) *Shared Mental Models*—members develop a shared understanding of what the common goal is and how each one is contributing to this goal [53]; and (iii) *Cognitive Centrality*—members who hold strong relevant expertise are influential; members of effective communities gradually move from being peripheral to becoming more central and engaged in the community [54]. It has been shown that community patterns based on these processes can be derived from the community graph [42, 44].

Interaction activities (e.g. communication and argumentation) are crucial for collaboration and decision making. Hence, in Dicode, we envisaged a novel mechanism for modeling relationships between content and people based on interaction data—both interaction with content (e.g. a medical image) or argumentative discussions between members. Input for the community modeling mechanism can be interaction log data, including provenance data of collaboration and decision making activities, as well as engagement in dialogue/argumentative interactions. We also expect that certain patterns, related to collaboration and decision making, will be detected by analyzing the community interaction data. Finally, Dicode can take advantage of the ontology that can be used to relate people and content.

2.3.4.7 Approaches for Community Support

There is a growing interest in providing intelligent support for teams, groups and communities. Visualization techniques are among the most popular methods that can be employed to present group and community models in a graphical way, to help groups function more effectively [50, 56], to motivate community participation [39], and to make members aware of reciprocal relationships [57]. The key limitation of visualization techniques is their passive influence on the functioning of the community, e.g. while examining graphical representations members may not be able to see how their contribution could be beneficial for the community as a whole and what activities they can engage in.

Different tools and algorithms have been developed to support people in locating expertise on a specific subject inside groups or VCs [48, 58]. There is a growing body of research on intelligent group/community interventions, e.g. notification [59], feedback [60], or promotion of cognitively central members [40, 61]. Community interventions aiming at improving the functioning of the community as an entity are presented in [62]. This includes pointing at connections between members which have not been exploited or encouraging cognitively central and peripheral members to engage in interactions beneficial for the whole community.

2.3.4.8 Implications for Dicode

Dicode required a new approach to support collaborative teams. On the one hand, teams can be supported to better conduct internal activities linked to collaboration and decision making (e.g. consider all possible aspects when making a judgment, compare opinions from diverse sources). On the other hand, teams can be made aware of external processes related to their decision making process, such as trends (e.g. specific data sources can be used for specific purposes), influences (e.g. patients' attitude to a drug may be influenced by the overall opinion of this drug or similar ones in public forum), reputation and trust (e.g. specific data or sources can have higher reputation among scientists).

2.4 Integration Issues

Software systems are usually designed to work in isolation. During the last years, new requirements and challenges have appeared due to the evolution and improvement of the communication networks. At present, systems frequently need to exchange heterogeneous data and collaborate with other applications. However, integration is a complex problem depending on many factors such as system architectures, operating systems, type of the components and information to be integrated, coupling and use of the systems, performance requirements, data heterogeneity and semantics, user interfaces, middleware, and availability of resources [63]. In the following, we analyze the state-of-the-art on integration issues from two different points of view, namely data integration (Sect. 4.1) and applications integration (Sect. 4.2).

2.4.1 Data Integration

Nowadays, it is difficult to imagine a modern organization, company or institution storing all their data in only one system. Usually, they have the information distributed among several physical devices, i.e. computers, hard disks, databases, CD/DVD, etc. Efficient integration of such information becomes crucial in order to perform analysis, experiments and decision-making tasks. The underlying idea of data integration is simple: an organization has interrelated information in different places and it wants to retrieve all that information in a uniform way just making a unique query. But such simplicity is far from reality. Data integration has to consider a series of issues ranging from technical ones (e.g. computer features, database management systems, communication parameters) to problems related to the representation of information (e.g. information coding, representation models, data heterogeneity).

Traditionally, two major approaches to integrate data are considered: *centralized* versus *distributed/federated* ones. In the literature, such approaches are also known as *data translation* and *query translation*, respectively. The main difference between them essentially lies in the physical place where data is stored and the methods and technologies used to retrieve such data. Apart from these approaches, we can consider another data integration approach called *information linkage*. It is closely related to the web environment where information is integrated using static hyperlinks [64]. But this approach does not constitute real integration. It is just collections of links regarding web pages about the same topic. Examples of this type of integration are MEDLINE [65], PDB [66] or Prosite [67].

2.4.1.1 Centralized Approaches

The most representative example of centralized approaches is the *data warehouse*. A data warehouse is a database management system which gathers data coming from several databases. Data are imported into the data warehouse using a common format. To carry out this task, data needs to be transformed from the original format to the new one. Such transformation process is performed by an entity/program called ETL (Extract, Transform and Load). These processes are usually executed in background mode. A typical data warehouse architecture can be seen in Fig. 2.1.

The main characteristic of data warehouses is that data are physically stored in a common database. Performance and efficiency constitutes the major benefit of centralized approach. The major drawback lies in the potential size of the data warehouse especially when many data sources are integrated. Another non-trivial problem is how to keep continuously updated the central repository. The centralized approach is particularly suitable for systems which do not change frequently. Many examples of data warehouses systems in the biomedical area can be found in the literature [68–72].

2.4.1.2 Distributed/Federated Approaches

Compared to centralized approaches, the main characteristic of distributed systems is that data remain physically stored in their original databases. Each time some information is required, data is directly retrieved from the sources. This task is usually carried out by a middleware or mediator layer, which deals with syntactic and semantic heterogeneity of the data and the data sources. Whenever a query is launched, the mediator decomposes it and sends the appropriate sub-queries to all

Fig. 2.1 Simplified data
warehouse architecture

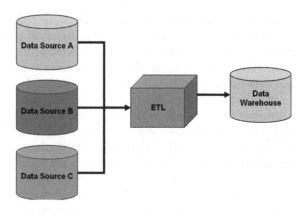

Fig. 2.2 Simplified
architecture of a distributed
system

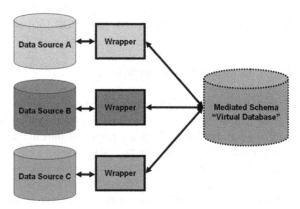

databases which are affected by the original query. Such decomposition is performed according to an existing global schema and the relations established between the global schema and the databases. Such relations are called mappings. Generation of sub-queries requires some transformations of the original query to be understood by the underlying databases. In some approaches, there are a kind of adapters, called wrappers, which facilitate the communication between the mediator and the databases. Finally, the mediator collects all results, unifies them and returns an integrated result. Figure 2.2 shows a simplified schema of a distributed system.

This kind of systems can be highly complex. Dependency from communication networks and access times are the main drawbacks of this approach. According to [64, 73], query translation approaches can be classified into four categories:

- **Pure Mediation**. The key components of these systems are the *wrappers*. Each database to be integrated into the global system needs a different *wrapper*, while such *wrappers* are completely different among them. Representative examples of pure mediation systems are TSIMMIS [74], DISCO [75], DIOM [76], HERMES [77] and BioDataServer [78].

- **Global as View (GAV)**. These systems define a global conceptual schema from the particular schemas of the databases integrated, i.e. the global schema is the combination of all database schemas. Once the global schema is created, a mapping has to be established between the global schema and each database. The main benefit of this approach is that the global schema describes very well the underlying databases and a common vocabulary is shared by all databases. On the other hand, any change in the structure of the databases or adding new databases to the system force one to reconsider the whole global schema. Examples of systems following this approach are SIMS [79], Ariadne [80] and TAMBIS [81].
- **Local as View (LAV)**. Systems following the LAV approach define multiple conceptual schemas, one per database integrated. They also have a global schema, but each conceptual schema (mapping) is expressed according to the global schema. In this case, a common vocabulary is not shared by all databases. Accuracy and reliability of systems depend on how good the global schema is (in terms of representing the contents of the databases). This approach enables a high extensibility and modularity but queries require more complex processing. A system following this approach is OBSERVER [82].
- **Hybrid approaches**. This approach tries to combine and take advantage of the benefits described in the previous approaches. Multiple particular schemas are created according to a global schema but using a common vocabulary shared by all. Systems following this approach include SEMEDA [83] and ONTOFUSION [84].

2.4.2 Applications Integration

There exist thousands of applications and services already developed around the world. This number grows exponentially. Software integration architectures enable reusing existing applications and services to work together with new developments. Reusing applications has many benefits, from reducing cost to shorten the development time.

There are several architectural styles for software development according to different aspects such as communication, domain, relationship, structure, data, data flows or object oriented. Due to the nature of the Dicode project, our analysis was focused on distributed architectures. The main technologies are:

- Remote Procedure Call (RPC);
- Remote Method Invocation (RMI);
- Common Object Request Broker Architecture (CORBA);
- Event-based Architectures;
- Service-Oriented Architectures (SOA), and
- Resource-based Architectures (REST).

In the next sections, SOA and REST are analyzed in more detail since they constitute the current state-of-the-art in integration technologies.

2.4.2.1 Service-Oriented Architecture

Software systems require a proper integration between their modules and components. To ensure a successful integration, there are two factors that must be taken into account: coupling and adaptation to standards. These factors may hamper the tasks of designing and implementing software systems, converting them into products that do not scale and hence will not be used.

The coupling of a system is given by the degree of interdependency among modules and programs. It is desirable that this interdependency remains as little as possible because a loose coupling between components facilitates the modification of any of the modules without affecting the rest of the parties. If the modules are more dependent on each other, it is more complicated to integrate a module into another system without having to interact with all modules.

Adaptation to standards relies on the correct design and documentation of the system. Well-planned and designed systems have interfaces for integration between its modules. By using standards, the need to develop specific software to perform this integration is minimized.

Service-oriented architecture (SOA) is based on well-defined standards. It is focused on the low coupling between modules of the system. SOA is not a tool, technology or product, but a concept, a set of rules and principles to design software, regardless of the technology used during its development. SOA relies on the creation of some interfaces that abstract away its underlying complexity. By using such interfaces, clients and providers may establish communications, just knowing the inputs and outputs of the services.

SOA is usually implemented using *web services*. A web service is a set of standards and protocols which allow the information exchange between different applications. They provide low coupling and adaptation to the standards demanded by SOA. Due to its nature, web services are especially appropriated to implement SOA standards. However, we could apply SOA without web services. Applications using SOA can be found in distributed environments. They communicate with each other through the interfaces to obtain information or execute a particular workflow. These applications are platform-independent, i.e. they can be developed with different tools, languages and platforms. Standards followed by web services are SOAP, XML, WSDL and UDDI.

Examples of SOA technologies in the field of Bioinformatics are the National Center for Biotechnology Information (NCBI) website, the European Bioinformatics Institute (EBI) services, BioMart (http://www.biomart.org), the CaBIG initiative (www.cabig.cni.nih.gov), myGrid project (www.mygrid.org) and Bio-MOBY (www.biomoby.org).

2.4.2.2 Resource Oriented Architecture: REST

REST is an acronym that stands for REpresentational State Transfer [85]. REST is an architecture style based on the resources that are provided by the World Wide

Web. Roy Fielding developed these concepts in his Ph.D. thesis titled "Architectural Styles and the Design of Network-based Software Architectures" (defended in the University of California, Irvine, 2000).

REST contributes with a set of constraints for designing network architectures. Thus, REST is not an exhaustive standard applied to network applications but an architectural style that is supported by existing standards such as URI, HTTP and HTML. Indeed, Roy Fielding was coauthor of the HTTP specification while working with the W3C in the definition of standards for the World Wide Web. The idea of Fielding was to represent the behavior of a web application that could be taken as a model because of its design. In this context, the web applications that caught the interest of Fielding were those that fit the model of distributed hypermedia system: any kind of resources (text, image, video, etc.) linked to each other through hyperlinks and placed or distributed in different servers over the network. REST is based on the HTTP, URI, MIME, XML and WADL standards [86–88].

2.4.2.3 Comparison: SOA Versus REST

REST appeared as an evolution of SOA to solve some problems of the latter in certain projects. REST implements web services ignoring the SOAP protocol. REST gives more importance to information while SOAP is opting for message exchange. A comparison between web services based on SOAP and REST is presented below [89].

- **Technology**. While SOAP offers a multitude of operations that operate on limited resources, REST instead offers a few operations that can interact with a multitude of resources. In the case of SOAP, these operations are given by a flow of events. In the case of REST, they interact directly with users via forms. REST has a consistent mechanism for naming resources through URI, whereas SOAP has not.
- **Protocol**. SOAP XML documents are strongly typed based on XML Schema, while REST uses a self-describing XML. Although both technologies use HTTP as transport protocol, SOAP might use another one. HTTP is also used by REST as application protocol. Finally, REST works synchronously, while SOAP can work on both a synchronous or asynchronous mode.
- **Service Description**. SOAP is based on contracts using the WSDL standard, while REST establishes a set of user-oriented documents defining the directions of requests and responses. WSDL documents (from SOAP) are more difficult to be understood by humans compared to REST definitions. Besides, WSDL allows building automatically web services clients from the descriptions. Since November 2006, REST includes the WADL standard to emulate SOAP mechanisms.
- **State management**. Both approaches enable state management but use different methods. Since REST servers are stateless, i.e. servers do not remember previous queries or invocations, each request must contain all the information

needed to answer the petition. State transitions can be simulated using cookies in the client side and incorporating extra data and links to resources through the URIs or in the message payload. On the other hand, SOAP can maintain the current state of the services on the servers using sessions, although SOAP session headers are not standard. Additionally, SOAP may communicate such information using message payload or WS-Addressing specification.

- **Security**. SOAP provides security mainly through the WS-Security protocol. This protocol defines how to provide integrity and confidentiality on exchanged messages and how to attach security tokens, such as SAML, Kerberos and X.509 certificates. It is associated with other specifications such as WS-Policy, WS-Trust or WS-SecureConversation, WS-Federation, WS-Privacy and WS-Test. Security in REST is based on the mature HTTPS protocol to provide point-to-point SSL secure communications. This protocol can be also used in SOAP as complement of the WS-Security standards. REST does not have defined open standards comparable to WS-Security for distributed transactions. Therefore, the use of proprietary security implementations is needed when HTTPS is not enough to provide digital signatures or detailed authentication and authorization.

- **Design methodology**. When designing applications, it is important to consider that REST is focused on resources, while SOA is focused on message exchange. In case of REST, we have to consider what information and resources will be available as services, while in SOA we have to identify the operations appearing in the WSDL document. In REST, it is necessary to define different URLs to address the web services; in SOA, we must define a data model for the content of messages. Finally, in SOA we need to implement, register and deploy the web services, while in REST it is only needed to implement and deploy the web service without any registration.

2.5 Conclusions

Aiming to provide useful insights into the exploitation and advancement of existing collaboration and decision making support technologies, this chapter has reviewed the state-of-the-art on related approaches. The major dimensions considered concern the issues of information overload and cognitive complexity. In addition, the chapter has reviewed the state-of-the-art on integration issues from two different aspects, namely data integration and applications integration.

The Dicode project has exploited and significantly advanced the state-of-the-art in the directions that have been elaborated in this chapter. The project developed efficient and dependable services that augment problem solving, sense making and decision making support for critical, information-bound domains in which our ability to share and exploit information is surpassed by the rate of its expansion in both size and complexity (see Chaps. 4, 5 and 6). The abovementioned augmentation resulted out of the meaningful and efficient integration and orchestration of all Dicode services, which had also to pay attention on scalability, flexibility and performance issues (see Chap. 7).

References

1. Economist: A special report on managing information: data, data everywhere. Economist (2010)
2. Schick, A.: Information overload: a temporal approach. Acc. Organ. Soc. **15**(3), 199–220 (1990)
3. Kirsh, D.: A few thoughts on cognitive overload. Intellectica **1**(30), 19–51 (2000)
4. Batra, D.: Cognitive complexity in data modeling: causes and recommendations. Requirements Eng. **12**, 231–244 (2007)
5. Thompson, J.D.: Organizations in Action. McGraw-Hill, New York (1967)
6. Gorry, G.A., Scott Morton, M.: A framework for management information systems. Sloan Manage. Rev. **13**(1), 50–70 (1971)
7. Pearson, J.M., Shim, J.P.: An empirical investigation into DSS structures and environments. Decis. Support Syst. **13**, 141–158 (1995)
8. Forgionne, G., Gupta, J., Mora, M.: Decision making support systems: achievements, challenges and opportunities. In: Mora, M., Forgionne, G., Gupta, J. (eds.) Decision Making Support Systems: Achievements and Challenges for the New Decade, pp. 392–403. Idea Group, Hershey (2002)
9. Shim, J.P., Warkentin, M., Courtney, J.F., Power, D.J., Sharda, R., Carlsson, C.: Past, present and future of decision support technology. Decis. Support Syst. **33**, 111–126 (2002)
10. DeSanctis, G., Gallupe, R.B.: A foundation for the study of group decision support systems. Manage. Sci. **33**(5), 589–609 (1987)
11. Arnott, D., Pervan, G.: A critical analysis of decision support systems research. J. Inf. Technol. **20**(2), 67–87 (2005)
12. Courtney, J.: Decision making and knowledge management in inquiring organizations: toward a new decision making paradigm for DSS. Decis. Support Syst. **31**, 17–38 (2001)
13. Alter, S.: Why persist with DSS when the real issue is improving decision making? In: Jelassi, T., Klein, M.R., Mayon-White, W.M. (eds.) Decision Support Systems: Experiences and Expectations, IFIP, pp. 1–11. North-Holland, Amsterdam (1992)
14. Angehrn, A., Jelassi, T.: DSS research and practice in perspective. Decis. Support Syst. **12**, 267–275 (1994)
15. Turban, E., Aronson, J.E.: Decision Support Systems and Intelligent Systems, 6th edn. Prentice Hall, Upper Saddle River (2001)
16. O'Brien, James, Marakas, George: Introduction to Information Systems. McGraw-Hill, Inc., New York (2009)
17. Kimball, R., Ross, M.: The Data Warehouse Toolkit: The Complete Guide to Dimensional Modeling, 2nd edn. Wiley, New York (2002)
18. Thomsen, E.: OLAP Solutions: Building Multidimensional Information Systems, 2nd edn. Wiley, New York (2002)
19. Fayyad, U., Piatetsky-Shapiro, G., Smyth, P.: From data mining to knowledge discovery in databases. AI Mag. **17**, 37–54 (1996)
20. Han, J., , KC-C.: Data Mining for Web Intelligence. Computer. **35**(11), 54–60 (2002)
21. Karacapilidis, N.: An overview of future challenges of decision support technologies. In: Gupta, J., Forgionne, G., Mora, M. (eds.) Intelligent Decision-Making Support Systems: Foundations, Applications and Challenges, pp. 385–399. Springer, London (2006)
22. Smoliar, S.W.: Interaction management: the next (and necessary) step beyond knowledge management. Bus. Process Manage. J. **9**(3), 337–352 (2003)
23. Prahalad, C.K., Hamel, G.: The core competence of the corporation. Harv. Bus. Rev. **68**(3), 79–91 (1990)
24. Nonaka, I.: A dynamic theory of organizational knowledge creation. Organ. Sci. **5**(1), 14–37 (1994)
25. Cohendet, P., Steinmueller, W.E.: The codification of knowledge: a conceptual and empirical exploration. Ind Corp. Change **9**(2), 195–209 (2000)

26. Holsapple, C.W., Whinston, A.B.: Decision Support Systems: A Knowledge-Based Approach. West Publishing Company, Saint Paul (1996)
27. Bhagat, J., Tanoh, F., Nzuobontane, E., Laurent, T., Orlowski, J., Roos, M., Wolstencroft, K., Aleksejevs, S., Stevens, R., Pettifer, S., Lopez, R., Goble, C.A.: BioCatalogue: a universal catalogue of web services for the life sciences. Nucleic Acids Res. **38**(Web Server issue), W689–W694 (2010)
28. Eemeren, F.H. van, Grootendorst, R., and Snoeck Henkemans, A.F. (eds.) Fundamentals of Argumentation Theory: A Handbook of Historical Backgrounds and Contemporary Developments. Lawrence Erlbaum Associates, Mahwah, NJ (1996)
29. Provis, C.: Negotiation, persuasion and argument. Argumentation **18**, 95–112 (2004)
30. Leuf, B., Cunningham, W.: The Wiki Way: Quick Collaboration on the Web. Addison-Wesley, Boston (2001). ISBN 0-201-71499-X
31. Hanneman A., Riddle M.: Introduction to social network methods. http://www.faculty.ucr.edu/hanneman/nettext/ (2005)
32. Wasserman, S., Faust, K.: Social Network Analysis: Methods and Applications. Cambridge University Press, Cambridge (1994)
33. Scheuer O., Loll F., Pinkwart N., McLaren M.B.: Computer-supported argumentation: a review of the state of the art. Int. J. Comput.-Support. Collab. Learn. **5**(1), 43–102 (2010)
34. Eppler, M.J., Mengis, J.: The concept of information overload: a review of literature from organization science, accounting, marketing, mis, and related disciplines. Inf. Soc. **20**, 325–344 (2004)
35. Palla, G., Barabási, A.L., Vicsek, T.: Quantifying social group evolution. Nature **446**(7136), 664–667 (2007)
36. Palla, G., Derefny, I., Farkas, I., Vicsek, T.: Uncovering the overlapping community structure of complex networks in nature and society. Nature **435**(9), 814–818 (2005)
37. Masthoff, J.: Group modeling: selecting a sequence of television items to suit a group of viewers. User Model. User-Adap. Inter. **14**(1), 37–85 (2004)
38. Lin, Y.-R., Chi, Y., Zhu, S., Sundaram, H., Tseng, B.: Facetnet: a framework for analyzing communities and their evolutions in dynamic networks. In: Proceeding of the 17th International Conference on World Wide Web, WWW '08. ACM, Beijing, China (2008)
39. Cheng, R., Vassileva, J.: Design and evaluation of an adaptive incentive mechanism for sustained educational online communities. J. User Model. User Adap. Inter. **V16**(3), 321–348 (2006)
40. Bretzke, H., Vassileva, J.: Motivating cooperation on peer to peer networks. In: 9th International Conference on User Modelling 2003. Springer, Berlin (2003)
41. Kleanthous, S., Dimitrova, V.: Modelling semantic relationships and centrality to facilitate community knowledge sharing. In: Proceedings of the 5th International Conference on Adaptive Hypermedia and Adaptive Web-Based Systems (AH'08). Springer, Berlin (2008)
42. Kleanthous, S., Dimitrova, V.: Analyzing community knowledge sharing behavior, user modeling. In: Adaptation, and Personalization. Springer, Berlin (2010)
43. Davies, J., Duke, A., Sure, Y.: OntoShare: a knowledge management environment for virtual communities of practice. In: Proceedings of the International Conference on Knowledge Capture, K-CAP '03. ACM, Sanibel Island, FL, USA (2003)
44. Kleanthous, S., Dimitrova, V.: Detecting changes over time in a knowledge sharing community. In: Proceedings of the 2009 IEEE/WIC/ACM International Joint Conference on WI and IAT. IEEE Computer Society, Washington, DC, USA (2009)
45. Song, X., Tseng, B., Lin, C.-Y. and Sun, M.-T. (2005): ExpertiseNet: relational and evolutionary expert modeling. In: Lecture Notes in Computer Science, vol. 3538. Springer, Heidelberg
46. Fu, Y., Xiang, R., Liu, Y., Zhang, M., Ma, S.: Finding experts using social network analysis. In: Proceedings of the IEEE/WIC/ACM International Conference on Web Intelligence, WI '07. IEEE Computer Society, Silicon Valley, CA, USA (2007)
47. Lin, Y.-R., Sundaram, H., Chi, Y., Tatemura, J., Tseng, B.: Blog community discovery and evolution based on mutual awareness expansion. In: Proceedings of the IEEE/WIC/ACM

International Conference on Web Intelligence, WI '07. IEEE Computer Society, Silicon Valley, CA, USA (2007)

48. Zhang, J., Ackerman, M., Adamic, L.: Expertise networks in online communities: structure and algorithms. In: International Conference on WWW 2007. ACM, Alberta, Canada (2007)

49. Hubscher, R., Puntambekar, S.: Modeling learners as individuals and as groups. In: Adaptive Hypermedia and Adaptive Web-Based Systems. pp. 300–303 (2004)

50. Kay, J., Maisonneuve, N., Yacef, K., Reimann, P.: The big five and visualisations of team work activity. In: Intelligent Tutoring Systems 2006. Lecture Notes in Computer Science, vol. 4053/2006. Springer, Berlin (2006)

51. Falkowski, T., Barth, A., Spiliopoulou, M.: DENGRAPH: a density-based community detection algorithm. In: IEEE/WIC/ACM International Conference on Web Intelligence. IEEE Computer Society, Silicon Valley (2007)

52. Falkowski, T., Spiliopoulou, M.: Users in volatile communities: studying active participation and community evolution. In: Proceedings of the International Conference on User Modeling 2007. LNCS, vol. 4511/2007. Springer, Berlin (2007)

53. Mohammed, S., Dumville, B.C.: Team mental models in a team knowledge framework: expanding theory and measurement across disciplinary boundaries. J. Organ. Behav. **22**(2), 89–106 (2001)

54. Ilgen, D.R., et al.: Teams in Organizations: from input-process-output models to IMOI models. Annu. Rev. Psychol.(56), 517–543 (2005)

55. Wenger, E.: Communities of practice: learning as a social system. Syst. Thinker **9**(5), 2–3 (1998)

56. Upton, K., Kay, J.: Narcissus: group and individual models to support small group work. In: Proceedings of the 17th International Conference on UMAP. Springer, Trento (2009)

57. Sankaranarayanan, K., Vassileva, J.: Visualizing reciprocal and non-reciprocal relationships in an online community. In: Proceedings of International Workshop on Adap-tation and Personalization for Web 2.0 (AP-Web 2.0 2009). UMAP'09, CEUR Workshop Proceedings, Trento, Italy (2009)

58. Shami, S., Yuan, C., Cosley, D., Xia, L., Gay, G.: That's what friends are for: facilitating 'who knows what' across group boundaries. In: Proceedings of the ACM 2007 GROUP Conference. ACM, FL, USA (2007)

59. Ardissono, L., Bosio, G.G.A., Petrone, G.: Context-aware notification management in an integrated collaborative environment. In: Proceedings of International Workshop on Adaptation and Personalization for Web 2.0 (AP-Web 2.0 2009). UMAP'09, Trento, Italy (2009)

60. Baghaei, N., Mitrovic, T.: From modelling domain knowledge to metacognitive skills: extending a constraint-based tutoring system to support collaboration. In: International Conference on User Modeling 2007. Springer, Corfu (2007)

61. Farzan, R., DiMicco, J., Brownholtz, B.: Spreading the honey: a system for maintaining an online community. In: Proceedings of the ACM GROUP 2009 Conference. ACM, FL, USA (2009)

62. Kleanthous, S.: Intelligent support for knowledge sharing in virtual communities. Ph.D. thesis, University of Leeds, UK (2010)

63. Ziegler, P. and Dittrich, K.R.: Three decades of data integration—all problems solved? In: 18th IFIP World Computer Congress (WCC 2004), vol. 12. Building the Information Society, vol. 2004, pp. 3–12 (2004)

64. Anguita, A., Martin, L., Perez-Rey, D., Maojo, V.: A review of methods and tools for database integration in biomedicine. Curr. Bioinform. **5**(4), 253–269 (2010)

65. Lindberg, C.: The unified medical language system (UMLS) of the national library of medicine. J. Am. Med. Rec. Assoc. **61**(5), 40–42 (1990)

66. Berman, H.M., Westbrook, J., Feng, Z., Gilliland, G., Bhat, T.N., Weissig, H., Shindyalov, I.N., Bourne, P.E.: The protein data bank. Nucleic Acids Res. **28**(1), 235–242 (2000)

67. Hulo, N., Bairoch, A., Bulliard, V., Cerutti, L., De Castro, E., Langendijk-Genevaux, P.S., Pagni, M., Sigrist, C.J.: The PROSITE database. Nucleic Acids Res. **34**(Database issue) (2006)
68. Etzold, T., Argos, P.: SRS-an indexing and retrieval tool for flat file data libraries. Comput. Appl. Biosci. CABIOS **9**(1), 49–57 (1993)
69. Maglott, D., Ostell, J., Pruitt, K.D., Tatusova, T.: Entrez Gene: gene-centered information at NCBI. Nucleic Acids Res. **35**(Database issue), D26–D31 (2007)
70. Shah, S.P., Huang, Y., Xu, T., Yuen, M.M., Ling, J., Ouellette, B.F.: Atlas—a data warehouse for integrative bioinformatics. BMC Bioinform. **6**(1), 34+ (2005)
71. Lee, T.B., Fielding, R., Masinter, L.: Uniform resource identifier (URI): generic syntax. RFC 3986 (2005)
72. Zhu, X., Wang, J.: MyBASE: a database for genome polymorphism and gene function studies of Mycobacterium. BMC Microbiol. **9**, 40+ (2009)
73. Sujansky, W.: Heterogeneous database integration in biomedicine. J. Biomed. Inform. **34**(4), 285–298 (2001)
74. Molina, H.G., Papakonstantinou, Y., Quass, D., Rajaraman, A., Sagiv, Y., Ullman, J.D., Vassalos, V., Widom, J.: The TSIMMIS approach to mediation: data models and languages. J. Intel. Inf. Syst. **8**(2), 117–132 (1997)
75. Tomasic, A., Raschid, L., Valduriez, P.: Scaling access to heterogeneous data sources with DISCO. IEEE Trans. Knowl. Data Eng. **10**(5), 808–823 (1998)
76. Liu, L., Pu, C.: An adaptive object-oriented approach to integration and access of heterogeneous information sources. Distrib. Parallel Databases **5**, 167–205 (1997)
77. Adali, S., Candan, K.S., Papakonstantinou, Y., Subrahmanian, V.S.: Query caching and optimization in distributed mediator systems. SIGMOD Rec. **25**(2), 137–146 (1996)
78. Freier, A., Hofestädt, R., Lange, M., Scholz, U., Stephanik, A.: BioDataServer: a SQL-based service for the online integration of life science data. Silicon Biol. **2**(2), 37–57 (2002)
79. Arens, Y., Chee, C.Y., Hsu, C.N., Knoblock, C.A.: Retrieving and integrating data from multiple information sources. Int. J. Coop. Inf. Syst. **2**(2), 127–158 (1993)
80. Knoblock, C.A., Minton, S., Ambite, J.L., et al.: The Ariadne approach to web-based information integration. Int. J. Coop. Inf. Syst. **10**(1–2), 145–169 (2001)
81. Stevens, R., Baker, P., Bechhofer, S., Ng, G., Jacoby, A., Paton, N.W., Goble, C.A., Brass, A.: TAMBIS: transparent access to multiple bioinformatics information sources. Bioinformatics **16**(2), 184–185 (2000). (Oxford, England)
82. Mena, E., Kashyap, V., Sheth, A.P., Illarramendi, A.: OBSERVER: an approach for query processing in global information systems based on interoperation across pre-existing ontologies. In: Conference on Cooperative Information Systems, pp. 14–25 (1996)
83. Köhler, J., Philippi, S., Lange, M.: SEMEDA: ontology based semantic integration of biological databases. Bioinformatics **19**(18), 2420–2427 (2003)
84. Pérez-Rey, D., Maojo, V., García-Remesal, M., Alonso-Calvo, R., Billhardt, H., Martin-Sánchez, F., Sousa, A.: ONTOFUSION: ontology-based integration of genomic and clinical databases. Comput. Biol. Med. **36**(7–8), 712–730 (2006)
85. Fielding, R.T.: Architectural styles and the design of network-based software architectures. Ph.D. thesis (2000)
86. Hadley, M.J.: Web application description language (WADL). Technical report, Mountain View, CA, USA (2006)
87. Chinnici, R., Moreau, J.-J., Ryman, A., Weerawarana, S.: Web services description language (WSDL) Version 2.0 Part 1: core language. Technical report (2007)
88. Mandel, L.: Describe REST Web services with WSDL 2.0 (2008)
89. Pautasso, C., Zimmermann, O., Leymann, F.: Restful web services vs. "big" web services: making the right architectural decision. In: Proceeding of the 17th International Conference on World Wide Web, WWW '08, pp. 805–814. ACM, New York, USA (2008)

Chapter 3
Requirements for Big Data Analytics Supporting Decision Making: A Sensemaking Perspective

Lydia Lau, Fan Yang-Turner and Nikos Karacapilidis

Abstract Big data analytics requires technologies to efficiently process large quantities of data. Moreover, especially in decision making, it not only requires individual intellectual capabilities in the analytical activities but also collective knowledge. Very often, people with diverse expert knowledge need to work together towards a meaningful interpretation of the associated results for new insight. Thus, a big data analysis infrastructure must both support technical innovation and effectively accommodate input from multiple human experts. In this chapter, we aim to advance our understanding on the synergy between human and machine intelligence in tackling big data analysis. Sensemaking models for big data analysis were explored and used to inform the development of a generic conceptual architecture as a means to frame the requirements of such an analysis and to position the role of both technology and human in this synergetic relationship. Two contrasting real-world use case studies were undertaken to test the applicability of the proposed architecture for the development of a supporting platform for big data analysis. Reflection on this outcome has further advanced our understanding on the complexity and the potential of individual and collaborative sensemaking models for big data analytics.

Keywords Requirement elicitation · Model-driven · Sensemaking conceptual architecture

L. Lau (✉) · F. Yang-Turner
School of Computing, University of Leeds, Leeds LS2 9JT, UK
e-mail: L.M.S.Lau@leeds.ac.uk

F. Yang-Turner
e-mail: F.Yang-Turner@leeds.ac.uk

N. Karacapilidis
University of Patras and Computer Technology Institute & Press "Diophantus",
26504 Rio Patras, Greece
e-mail: nikos@mech.upatras.gr

N. Karacapilidis (ed.), *Mastering Data-Intensive Collaboration and Decision Making*,   49
Studies in Big Data 5, DOI: 10.1007/978-3-319-02612-1_3,
© Springer International Publishing Switzerland 2014

3.1 Introduction

The "big data" phenomenon is now present in every sector and function of the global economy [29]. Contemporary collaboration settings are often associated with huge, ever-increasing amount of multiple types of data, which vary in terms of relevance, subjectivity and importance. Extracted knowledge may range from individual opinions to broadly accepted practices. Today's businesses face challenges not only in data management but in big data analysis, which requires new approaches to obtain insights from highly detailed, contextualised, and rich contents. In such settings, collaborative sensemaking very often take place, orchestrated or otherwise, prior to actions or decision making [34]. However, our understanding on how these tools may interact with users to foster and exploit a synergy between human and machine intelligence quite often lags behind the technologies.

The term "data analytics" is often used to cover any data-driven decision making. A major investment in big data, properly directed, can result not only in major scientific advances, but also lay the foundation for the next generation of advances in science, medicine, and business [1]. To help decision making, data analysts choose informative metrics that can be computed from available data with the necessary algorithms or tools, and report the results in a way the decision makers can comprehend and act upon. Big data analytics is a workflow that distils terabytes of low-value data (e.g., every tweet) down to, in some cases, a single bit of high-value data (e.g., should Company X acquire Company Y?) [5].

Technologies such as data mining, machine learning and semantic web are being exploited to build infrastructures and advanced algorithms or services for big data analytics. Most of the services and algorithms are built in a technology-driven manner with little input from users to drive the development of the solutions. This may be due to: (1) users usually have few ideas about how the emerging technologies can support them; (2) problems described by users are quite general, such as "information overload", "data silos everywhere" or "lack of holistic view", and (3) goals set by users are often unclear, such as "find something valuable", "get an impression", or "obtain deep understandings". It is challenging to follow traditional approach of gathering user requirements to lead solution development using emerging technologies [16].

Another approach could be a technology-driven one, i.e., how to make the technology improve user's work practice. However, given a diverse set of business analytics situation and the fact that more and more analytics algorithms are developed, it is challenging to leverage the strengths and limitations of Big Data technologies and apply them in different domains [15].

This chapter sets out to bridge the gap between user-driven and technology-driven approaches for requirements analysis in big data problems and addresses the following research questions:

- *Question 1*: *How to derive requirements in big data analytics which are drawn from user sensemaking behaviour?*

- ***Question 2***: *Can we extract commonalities and differences across diverse application domains to advance our understanding of requirements for big data analytics?*
- ***Question 3***: *Can a conceptual architecture be useful for bringing user and technology perspectives together to develop specific big data analytics platform?*

Led by the above questions, we took a socio-technical approach on requirement modelling and adapted individual and collaborative sensemaking frameworks to guide our investigation on requirements of big data analytics. This study is part of the Dicode EU research project (http://dicode-project.eu), which aims at facilitating and augmenting collaboration and decision making in data-intensive and cognitively-complex settings. In particular, emphasis is given to the deepening of our insights about the proper exploitation of big data, as well as to collaboration and sensemaking support issues [9].

Our contribution is to operationalise sensemaking models to help understand the distribution of human and machine intelligence in the use of a big data analytics platform. The resulting conceptual architecture provides a framework which enables the main components to evolve systematically through a dialogue between users and technology suppliers.

The chapter proceeds as follows. In Sect. 3.2, we discuss sensemaking for big data analytics. In Sect. 3.3, we present our three-step methodology for requirement elicitation. In Sects. 3.4, 3.5 and 3.6, we describe the details of these three steps in the context of Dicode's use cases. In Sect. 3.7, we conclude the chapter and discuss on the implications of this study to support big data analytics.

3.2 Sensemaking for Big Data Analytics

Big data analytics, as an emerging area, has gained attention by both IT industry and academic research communities. From an infrastructure point of view, the top three commercial database suppliers—Oracle, IBM, and Microsoft—have all adopted Hadoop framework as their big data analytic platform [7]. Industry analysis pointed out that there are challenges not just in volume, but also in variety (the heterogeneity of data types, representation, and semantic interpretation) and velocity (both the rate at which data arrive and the time in which it must be acted upon) [6]. A community white paper developed by leading researchers across the United States argued that the challenges with big data include not just the obvious issues of scale, but also heterogeneity, timeliness, privacy and human collaboration [1]. This is a complex issue, and the gap between the number of companies which can make use of big data for transformational advantage and those that cannot is widening [9].

While smarter systems and algorithms may provide new perspectives into the data, humans are still indispensable in the analysis pipeline to turn them into information and knowledge. To analyse the data, an analyst may need to figure out questions suitable for the particular context, aiming to obtain new insight. In fact, we currently have a major bottleneck in the number of people empowered to ask questions of the data and analyse them [16]. As Barton and Court [2] aptly explained, a clear strategy for how to use big data analytics for competitive advantage requires a pragmatic approach to balance technical theories and practicalities. They suggested that business leaders can address short-term big data needs by working with their chief information officers to prioritize requirements.

In our study, we took a sensemaking perspective to understand the cognitive complexity of big data analytics, both individually and collaboratively. We then investigated the common activities of two use cases guided by the sensemaking frameworks to inform the design of a generic conceptual architecture for sensemaking. This architecture will illustrate the important components and their relationship at an abstract level for a quick overview of possible big data analytics solutions.

3.2.1 Individual Sensemaking

Sensemaking is an iterative cognitive process that the human performs in order to build up a representation of an information space that is useful to achieve his/her goal [25]. Sensemaking has been used in various fields such as organizational science [30], education and learning sciences [27], communications [4], human-computer interaction (HCI) [25], and information systems [26]. In communications, HCI and information science, sensemaking is broadly concerned with how a person understands and reacts to a particular situation in a given context. Cognitive models that describe the human sensemaking process can be helpful to point at what operations users in collaborative spaces may perform and what support they may need. One particular notional model developed by Pirolli and Card [22], which describes the sensemaking loop for intelligence analysis, helps us to identify particular sensemaking operations that a distributed data mining approach can support in a collaborative environment. The model distinguishes between two cognitive loops of intelligent analysis:

- The *foraging loop*, which involves operations such as seeking, searching, filtering, reading, and extracting information; and
- The *sensemaking loop*, which involves operations such as searching for evidence, searching for support, and re-evaluation, which aim to develop a mental model from the schema that best fits the evidence.

The operations involved in the defined loops highlight the importance of two high-level cognitive processes that a user of a collaborative space (e.g. discussion forum) performs: categorisation and schema induction [12]. In the foraging loop, the user tries to identify coherent categories, or topics, which summarise the underlying content and aid the user's filtering and searching to find the content relevant to the needs. In the sensemaking loop, on the other hand, the user tries to induce potential high-level schemas, or themes, from the identified topics. This is done by inducing the relations between the topics and evaluating the accuracy of those schemas. For example, if the user relates a collection of identified topics that include the terms {facebook, twitter, tweets, blogs, wordpress, wiki} to each other, she may be able to induce a high-level theme, which is {social media}, since the combination of the preceding topics is highly relevant to that theme.

Many forms of intelligence analysis are so-called sensemaking tasks [22]. Such tasks consist of information gathering, representation of the information in a schema that aids analysis, the development of insight through the manipulation of this representation, and the creation of some knowledge product or direct action based on the insight. The basis of an analyst's skill is to quickly organise the flood of incoming information and present his/her analysis in reports. The process of *creating a representation of a collection of information that allows the analyst to perceive structure, form and content within a given collection* is defined as sensemaking.

Different sensemaking models have revealed various characteristics of the analytical processes of intelligence analysts. Dervin illustrated that sensemaking occurs when a person embedded in a particular context and moving through time-space, experiences a gap in reality. Russell et al. [25] studied cost structure of sensemaking and modelled sensemaking as cyclic processes of searching for external representations and encoding information into these representations to reduce the cost of tasks to be performed. Klein et al. [13] defines sensemaking as a motivated, continuous effort to understand connections (which can be among people, places, and events) in order to anticipate their trajectories and act effectively.

3.2.2 Collaborative Sensemaking

Sensemaking extends beyond individuals making sense of their own information spaces. It is increasingly common for a group of people needing to work together to understand complex issues, combining information from multiple data sources and bringing together different experience and expertise towards a shared understanding.

However, there has been little exploration of how sensemaking takes place in collaborative work, let alone arriving at a unified view. Past studies reported sensemaking from different domains, perspectives or focuses. Ntuen [19] studied collaborative sensemaking in military coalition operations, where a group of people with different worldviews are collectively engaged in making sense of chaotic and ambiguous situations. Lee and Abrams [14] further explored sense-making regarding to collaboration which could entail innovation at two levels: joint learning in how to collaborate and coordinate work, and joint learning in how to represent and instantiate a design that does not yet exist. Qu and Hansen [24] proposed a conceptual model of collaborative sensemaking, which distinguishes between shared representation and shared understanding. They also argued that collaborators could develop a shared understanding by examining, manipulating and negotiating external representations. Paul and Reddy [20] have discussed a framework of collaborative sensemaking during Collaborative Information Seeking (CIS) activities and the design implications for supporting sensemaking in collaborative information retrieval tools.

3.3 A Model-Driven Requirement Elicitation Methodology

To answer the three research questions as discussed in the introduction, the following steps were taken in the big data analytics requirements methodology for Dicode across the use cases.

3.3.1 Context of Investigation and Use Cases

The Dicode project aimed at facilitating and augmenting collaboration and decision making in data-intensive and cognitively-complex settings. To do so, whenever appropriate, it built on prominent high-performance computing paradigms and large data processing technologies to meaningfully search, analyse and aggregate data existing in diverse, extremely large, and rapidly evolving sources. At the same time, particular emphasis was given to the deepening of our insights about the proper exploitation of big data, as well as to collaboration and sense making support issues. Building on current advancements, the solution provided by the Dicode project brings together the reasoning capabilities of both the machine and the humans. It can be viewed as an innovative "workbench" incorporating and orchestrating a set of interoperable services that reduce the data-intensiveness and complexity overload at critical decision points to a manageable level, thus permitting stakeholders to be more productive and effective in their work practices.

Two Dicode's use cases with different collaboration and decision making settings are used as illustration of our methodology in this chapter, each associated with diverse types of data and data sources.

- **Clinico-Genomic Research (CGR)**: this case concerns biomedical researchers who collaborate to explore scientific findings using very large datasets (a full description of this case appears in Chap. 8).
- **Social Opinion Monitoring (SOM)**: this case concerns social media marketing professionals who are frequently involved in strategic decisions about public presentation of branding, products or services (a full description of this case appears in Chap. 9).

3.3.2 Overview of the Methodology

The methodology deployed in the context of Dicode for requirement elicitation consists of the following three steps:

3.3.2.1 Step 1: Requirement Elicitation from Scenarios

A scenario-driven approach was used to capture from the stakeholders their views on current practice in selected data intensive and cognitively complex processes, and the initial vision on what could be improved from both users and technologists. A Dicode specific requirement elicitation strategy was designed and deployed to tackle the seemingly diverse use cases [31]. Common characteristics were extracted to identify common interests for technological innovation. This step mobilised ideas from both users and technologists.

3.3.2.2 Step 2: Application of Sensemaking Models

In addition to data collection from the ground, theoretical models for sensemaking were identified for a deeper understanding of sensemaking behaviour in each of the use cases. We considered an individual sensemaking model which provides a detailed view of data-driven analysis when trying to make sense of large volume of data. We supplemented it by a collaborative sensemaking model which presents the triggers of collaboration and characteristics of building shared understanding. The models provide a common framework for comparison in order to identify the commonalities and differences in sensemaking activities within different context. This step provided focus for users and technologists in positioning the benefits of proposed technical solutions and when these could be used.

3.3.2.3 Step 3: Conceptual Architecture for Big Data Analytics

Finally, a conceptual architecture was developed as a high level specification of how the various tools might work together for each of the use cases in a big data analytics platform. In designing the architecture, we followed the IS design research process proposed by Peffers and his colleagues [21] and aimed to create useful artefacts that solve relevant design problems in organizations [8, 18]. Usage scenarios were produced to walk through how the platform may be used. This step provided a high level blue print which could be used as a communication tool between the users and the technologists on requirements.

These steps are described in more detail in the following three sections of the chapter. In Sect. 3.4, we describe how the commonalities of the use cases were derived from both users and technologists. In Sect. 3.5, we present the underpinning sensemaking frameworks we adopted to guide our study: an intelligent analysis framework that presents how an individual analyst makes sense of large volume of data; and a framework of collaborative sensemaking during Collaborative Information Seeking (CIS) activities. We then introduce our generic conceptual architecture in Sect. 3.6 and its instantiations in two different application domains.

3.4 Requirement Elicitation from Scenarios

As the first step, we mobilised the tacit knowledge of use case partners by involving them in describing typical scenarios of current work practice in their areas. Data collection in this phase were directed at the facts about users and communities involved, data sources and data formats used as well as collaboration and decision making activities. Scenarios with sample data were provided on a wiki for all partners to read and discuss. These facts were essential to be understood as a benchmark so that Dicode could work on augmentation and facilitation to improve the current work practice.

A summary of two Dicode use cases (Table 3.1) shows that they had common issues related to a newly forming area for research, namely big data analytics.

> Fundamentally, big data analytics is a workflow that distils terabytes of low-value data (e.g., every tweet) down to, in some cases, a single bit of high-value data (should Company X acquire Company Y? can we reject the null hypothesis?) ([1], p. 50).

From a high level perspective, both use cases are from different domains; the users have different expertise and use different analytics tools. They deal with different data from different data sources, with stakeholders making different decisions for different purpose in their work. However, all of them are dealing with intelligent analysis to transform input data into knowledge product in order to see the "big picture" from a large collection of information.

Use case partners were then asked to describe their vision on a future system. It would be difficult for an individual user to give a complete scenario of how Dicode system might change their current practice. Therefore, in this phase, user stories were collected, in which users talked about their expectations about how Dicode could help (i.e. facilitate, augment) their work in the future. After analyzing all users' stories, we realized that at that stage users could only suggest small incremental change on their current work practice, which would not fully exploit the potentials that new technology would bring. In other words, information collected from users could not produce the desired innovation, and associated structured system requirements which could benefit from cutting edge technology. Input from technical partners into the requirement elicitation process was needed to stimulate a co-design culture.

We then encouraged the potential "sell and buy" within the Dicode project across all partners. This means effective communications between use case

Table 3.1 Summary of two dicode use cases

Use Cases	CGR	SOM
Application domain	Biomedical research	Social media marketing
Users	Biologists	Marketing analysts
	Biomedical researchers	Social media analysts
Expertise of users	Biology	Marketing
	Medical science, statistics	Communications
Analytics tools	Data collection, manipulation and analysis tools (such as R, or online data repositories)	Social media monitoring tools
Access of data sources	Public and private to research lab	Public
Input data	Gene-expression profiles (GEP)	News, blogs, tweets
	Phenotypic data	
	Molecular pathways (MP)	
	Annotation data	
Activities of intelligent analysis	Interpreting result	Formulating strategy
	Planning future research	Planning marketing campaign
Knowledge product	Scientific findings	Strategy for social media engagement
	Insights for experimental work (e.g. drug design)	

partners and technical partners about their ideas are facilitated. Technical partners were given the chance to evangelizing their ideas. For use case partners, they were encouraged to open their mind and seek new opportunities from new technologies. It was expected that use case partners and proposals from technical partners could diverge from each other. It is the future work practice, which takes both vision and proposals into account, to unify those differences. In Dicode, the results of this unifying were:

- a generic conceptual architecture to guide the design of services for each use case, in which requirements related to interfaces between services will be made explicit;
- a set of functional specifications that guide the first iteration of development.

3.5 Application of Sensemaking Models

To better understand the use cases in terms of intelligent analysis process, we included a social modelling approach to requirements engineering. This approach is driven by a priori understanding, through theories and models, of how human make sense of data and then apply that understanding to derive requirements from the use cases. Here, we describe the concepts and theoretical perspectives employed in our study, which are related to individual and collaborative sensemaking.

For individual sensemaking, we have chosen the model of Pirolli and Card [22] as it provides the means for identifying new technologies for improving the production

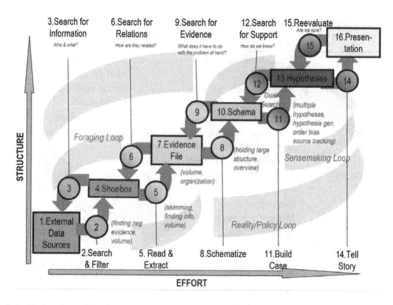

Fig. 3.1 Notional model of sensemaking loop for intelligence analysis derived from Cognitive task analysis (CTA) [22]

of new intelligence from massive data and its claim echoes ours in terms of sense-making is a process of transformation of information into a knowledge product.

Figure 3.1 summarizes how an analyst comes up with new information. The sequence of rectangular boxes represents an approximate data flow. The circles represent the process flow. The processes and data are arranged by degree of effort and degree of information structure. This is a process with lots of backward loops and seems to have one set of activities that cycle around finding information and another that cycles around making sense of the information, with plenty of interaction between these. The overall information processing can be driven by bottom-up processes (from data to theory) or top-down (from theory to data) and their analysis suggested that top-down process (process 2, 5, 8, 11, 14 in the diagram) and bottom-up processes (process 15, 12, 9, 6, 3) are invoked in an opportunistic mix. According to this framework, the processes of intelligent analysis of two Dicode use cases are identified in Table 3.2.

For collaborative sensemaking, we find Paul and Reddy's framework more relevant to our studies because it links individual sensemaking and collaborative sensemaking, and defines triggers and characteristics of sensemaking. In this framework, it highlights important factors that trigger collaborative sensemaking during a CIS activity, namely: ambiguity of information, role-based distribution of information, and lack of expertise. It shows that CIS activities are often initially split into tasks/sub-tasks and sub-tasks are performed by different group members, with different roles and expertise. Roles can be organisational or might be assigned informally. Within this context, action awareness information is shared amongst

Table 3.2 Processes of intelligent analysis of dicode use cases

Processes	CGR	SOM
(2) Search and filter	Extract/filter data of interests	Extract/filter data of interests
(3) Search for information	Search for complementary datasets	Search for relevant sources
(5) Read and extract	Extract patterns	Extract sentiments, opinions
(6) Search for relations	Search for similarities and differences among datasets	Search for trends
(8) Schematize	Biological interpretation the characteristics of data patterns	Create strategies, action plans
(9) Search for evidence	Produce or search for relevant datasets	Search for relevant events, influencers etc
(11) Build case	Create hypothesis	Create action plan
(12) Search for support	Consult the research community	Communicate with other parties
(14) Tell story	Produce scientific publication	Conduct marking activities
(15) Re-evaluate	Work on reviews of the publication	Evaluate the action result

group members even during individual sensemaking, i.e., group members keep each other aware of what they are doing.

The framework illustrated in Fig. 3.2 highlights that CIS activities often involve individual information seeking and sensemaking and then lead to collaboration. The framework lists some characteristics of collaborative sensemaking, namely, prioritising relevant information, sensemaking trajectories, and activity awareness. Prioritising the 'right' pieces of information as relevant enhances group sensemaking. Knowing the "path" that a group member followed to make sense of information helps other group members' sensemaking. Such paths are called sensemaking trajectories. Group members share and make sense of information, they create shared representations to store the information found and the sense made of that information. The characteristics and the triggers of collaborative sensemaking identified in this framework provide us a guideline to understand the demand of collaboration in Dicode use cases (Table 3.3).

3.6 Conceptual Architecture for Big Data Analytics

Derived from the Dicode use cases and sensemaking frameworks, we developed a generic conceptual architecture to support the characteristics (both differences and commonalities) of big data analytics. This conceptual architecture describes the important components and their relationship at an abstract level and provides a framework for specifying, comparing and contrasting big data analytics implementations.

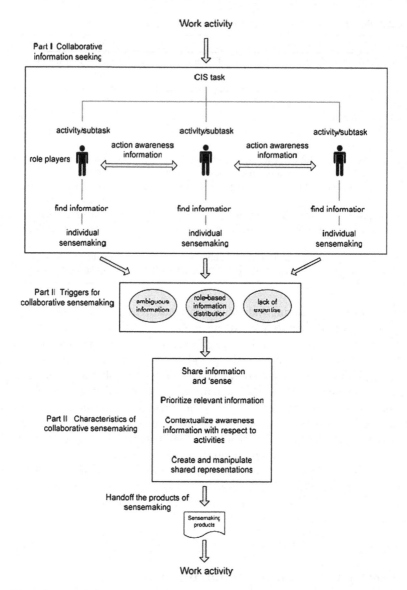

Fig. 3.2 A framework for collaborative sensemaking during Collaborative Information Seeking (CIS) activities [20]

The conceptual architecture aims to provide a framework without implementation of components, from which different big data analytics solutions can be constructed and implemented as long as they can fulfil their roles in the architecture. The conceptual architecture illustrates:

Table 3.3 Collaborative Sensemaking Triggers and Characteristics of Dicode Use Cases

	Description in the model	Examples in Dicode use cases	
Triggers	Ambiguous information	CGR	Acquire expert support
	Role-based information distribution		(e.g., a researcher needs the support of other researcher on whether his/her interpretation of the result is significant)
	Lack of expertise	SOM	Transfer knowledge to other parties for the result of social media analysis
Characteristics	Prioritizing relevant information	CGR	Get opinions from other scientists about choosing right datasets, databases or tools
	Sensemaking trajectories		
		SOM	Be aware of activities of other parties
	Activity awareness		Collaboratively transform data results to valuable insights

- A big data analytics solution consists of services or algorithms that exploit both machine capability (data-centric services) and human intelligence (collaboration-centric services).
- To facilitate and ensure the integration of machine capability and human intelligence, integration-centric services are needed to support users interact with both data-centric services and collaboration-centric services and provide mechanisms to integrate the result of two types of services.
- All services or algorithms together support the big data transformation from raw format to knowledge product (bottom-up) or from hypothesis to resources (top-down).
- Human intelligence should be involved in the whole process of data transformation, including configuring data-centric services, interpreting the result of data-centric services, collaborating with other experts on interpreting and sharing the results.

As shown in the architecture diagram (Fig. 3.3), there are three types of components:

- **Data-centric services**, which exploit large data processing technology to meaningfully search, analyse and aggregate data from heterogeneous data sources. The input of the data-centric services is structured and/or unstructured data from heterogeneous data sources. The output of data-centric services is searched or filtered information, discovered patterns or lists etc. The data-centric services aim to improve the processes of individual sensemaking.
- **Collaboration-centric services**, which support people and their interaction by capturing and sharing resources, opinions, arguments and comments among participants, so to facilitate the collective understanding of the issues related to data analysis. The input of the collaboration-centric services could be the output of data-centric services as well as the interactions (comments, arguments and discussions etc.) among all parties. The knowledge product (hypothesis, strategies

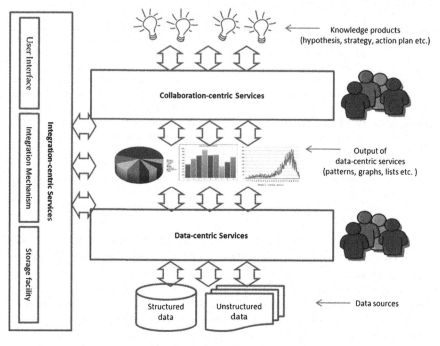

Fig. 3.3 Conceptual Architecture of Big Data Analytics

etc.) should be the outcome of their interaction. The collaboration-centric services aim to support collaborative sensemaking.

- **Integration-centric services**, which support data-centric services and collaboration centric-services. Integration-centric services are to ensure and facilitate the seamless integration of the independent services developed. Related functions include user interface, data storage and integration mechanisms etc. The integration-centric services implemented in Dicode project are the Dicode Workbench, the Dicode ONtology and the Storage Service.

The *Dicode Workbench* (see Chap. 7) provides a web user interface with functions of user management and service management. Through the Dicode Workbench, users can access different services (data-centric services and collaboration-centric services) developed within the Dicode project via widgets [3].

The *Dicode Ontology* (*DON*) is a multi-layered ontology, designed to address requirements from multiple use cases that involve sensemaking [28]. DON is used as a common vocabulary among services and service developers for enhancing the functionality of Dicode services. DON plays a crucial role to facilitate the integration and interoperability of services. The main idea is that some features of the services will be annotated using concepts included in the DON. The information about services and their annotations will be maintained in a central registry

(Dicode Service Registry—DSR). This registry will be available for the rest of the components of the Dicode environment through a REST interface.

The *Storage Service* is to provide Dicode users with a permanent and reliable storage place to keep resources accessible. The service will be as generic as possible to allow storing any kind of files (text plain, doc, pdf, html, xml, json, zip, etc.). The service provides mechanisms to upload files and retrieve them by using RESTful services. Additionally, meta-data information about files will be also stored to facilitate their search and location by search engines or services. These meta-data will contain information such as type of file (pdf, html, xml, etc.) or type of content (medical report, DNA sequence, etc.).

3.6.1 Usage Scenario for CGR

We present an example on how Bioinformatics researchers benefit from Dicode platform for their work:

Sarah (Ph.D. student), James (Postdoctoral Researcher) and John (Professor, supervisor of Sarah and James) are three researchers from a Breast Cancer research institution. They have conducted some studies on a small sample-size gene-expression microarray breast cancer dataset. The analysed result is not satisfactory but they believe that some extra datasets from public resources, such as Gene Expression Omnibus (GEO) with the same pathology characteristics can augment their sample size and allow them to identify some extra statistically significant genes.

*All of them are using **Dicode Workbench** to coordinate their work and support their research. Each of them has an account on the Dicode workbench and this enables them securely share their work. Using the **Storage service**, both Sarah and James have uploaded some graphs and data on what they have found out from their studies.*

*Working towards a publication, Sarah has added the **PubMed service** to their Dicode workbench. Using this service, she discovers relevant publications which address similar biological questions and may be used to justify their sample size choice. The result from PubMed tool has been recorded and can be seen by James and John at any time.*

*Having a brief idea about their sample size, the team "meet" in the **Collaboration workspace** to brainstorm their ideas and their opinions (agree, disagree, comments, ideas, support documents etc.).*

*To understand more about James' work, Sarah asks James to upload his R-script as she wants to know whether a few arguments (lines of code) could be rearranged. Using the **R service**, James run his R-script with some new arguments and a new graph is easily produced for everybody to assess the new strategy and decide on the significance of the results.*

*After a collaboration session, James has collected enough information about the data and sample size he needs for his task. James then launches the **GEO***

Recommender service to get the datasets. He types in the request describing the data and also the methodology he will apply. All qualified datasets are provided in a list.

From the list of recommended datasets, Sarah wants to find the functional interpretation of expressed genes in two datasets and compare them. She first launches the R service to identify expressed genes. In the second step, she uses Subgroup Discovery service, which provides a list of subgroups describing the expressed according to their molecular function and their role in biological process, which has shown a good match to their previous findings.

3.6.2 Architecture for CGR

Biomedical research has become increasingly interdisciplinary and collaborative in nature. The vast amount of the data available and the ever increasing specialised resources show that the way forward is to form biomedical research collaboration teams to address complex research questions. To support this use case, the Dicode solution (Fig. 3.4) is to support biomedical research community to work together dealing with increasing volume and diversity of data sources:

- Gene-Expression profiles (GEP): Gene-expression data (normalized or raw data);
- Phenotypic data: Supplementary, clinical or phenotypic data available;
- Molecular Pathways (MP): Data from known and established molecular networks;
- Annotation data: Reference databases for biomedical and genomic information.

The data-centric services are developed to deal with data processing and analysis in this field, such as:

- *Subgroup Discovery service* (see Chap. 5) provides the tool for the functional interpretation of gene expression data that combine and use knowledge stored in Gene Ontology database. The interpretation involves translating these data into useful biological knowledge. It is solved by constructing new features from Gene Ontology and finding the most interesting rules using Subgroup Discovery algorithm.
- *PubMed service* (see Chap. 8) provides access to PubMed but with extra improvements created for Dicode allowing data exchange with other services within Dicode workbench.
- *R service* (see Chap. 5) executes R-Scripts in Dicode and to perform custom data processing and data mining tasks.
- *GEO Recommender service* (see Chap. 5) provides relevant and interesting datasets from the Gene Expression Omnibus (GEO) repository according to users' preferences. The recommender service facilitates the reuse, retrieval and exchange of the GEO datasets by supporting the user in navigating in a large space of available datasets.

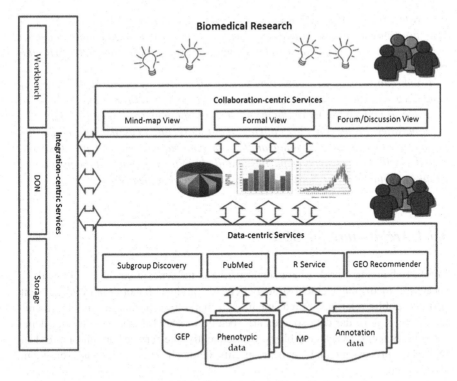

Fig. 3.4 Conceptual architecture for big data analytics in biomedical research

3.6.3 Usage Scenario for SOM

We present below an example on how social media analysts benefit from Dicode platform for their work:

A car manufacturer is launching a new product. In this process, three main parties are involved. One is a Brand Manager (Frank) from the marketing department of the company. The second one is a Social Media Analyst (Alice) working in a marketing consultancy. The third one is Social Media Engager (Natalie) working in a public relations agency responsible for social media engagement.

The Dicode Workbench allows all three parties to collaborate during the whole process. Frank has a question about first consumer experiences with the new product in the social web and gives a briefing to Alice.

Alice starts analysing the web and updates the results in the Collaboration workspace. She watches over social media and provides advice to the Brand Manager. She detects the significant conversations and news articles with the Topic Detection service and looks for insights as a basis for product development or communications from the blogs and tweets. If she wants to get deeper information on relevant tweets detected, she can use Keytrends service to show trends on Twitter, such as the top links for a certain day posted by twitter users. She can also use **Phrase Extraction service** *with the pre-trained sentiment model to monitor positive or negative sentiments that are expressed in connection with the brand.*

Frank can directly ask questions and/or give advice to control the research conducted by Alice. Natalie can access the results that Frank and Alice have provided to understand more about the current opinions from social media.

In parallel, Frank can start thinking about marketing activities to promote the product or to change packaging and/or communications. He can pre-align the activities with further involved parties in and out of the company. At the meantime, Frank can quickly brief Natalie on engaging with identified blogs.

3.6.4 Architecture for SOM

In a fast-changing world, where social media is influencing consumer demands, a successful media engagement strategy depends on the collaboration of all relevant parties—public relations, brand, media and marketing. In this instantiated architecture (Fig. 3.5), the data sources are specific in social media monitoring: dedicated news feeds, tweets and blogs. Consequently, the services are chosen to deal with data processing and analysis in this field, such as topic, and sentiment analysis, etc.

- *Named Entity service* (see Chap. 5) returns disambiguated Named Entities for Dicode's document corpora (currently Twitter and blogs). The service identifies Named entities of the following types: PERSON, PLACE, ORGANISATION and WORK and returns a Freebase URI for each entity. Named entity disambiguation is performed based on the context of the analysed surface form. The quality of disambiguation depends usually increases with text size.
- *Keytrends service* returns metadata about tweets on a selected day. Based on metadata: Hashtags (Top hashtags), Language (Languages of tweets), Country (Country code of Twitter user), Place (Places of Twitter user [only available for few tweets]) and Urls (Urls mentioned in tweets)
- *Topic Detection service* (see Chap. 5) gives the user a quick albeit superficial overview of the thematic content of a document collection, including a visualization of the results. The visualization provides a quick overview of the topics that are present in a text collection as well as their interrelations. Users will also be able to zoom in on a graph detail related to a particular topic.
- *Sentiment Analysis service* works on pre-trained models to extract positive and negative phrases from domain-specific text collections. It supports an interactive workflow, allowing the end-user train phrase extraction models interactively and apply them to a text collection.

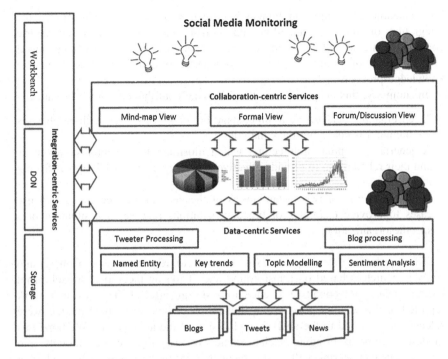

Fig. 3.5 Conceptual architecture for big data analytics in social media monitoring

3.7 Conclusion and Future Research

Traditionally, the task of the requirements analyst is to collect requirements and statements from stakeholders: the customer and representatives of users. These statements say what the system should do (functionality) and at what levels of quality (non-functional properties such as performance, reliability, extensibility, usability, and costs). However, users and customers are often not able to articulate these wants directly. Instead, the analyst needs to help them uncover their real needs. Users are often unaware of what is possible or have misconceptions about what is feasible, especially when technology is advancing quickly. For that, we claim that we should seek resources, such as existing models and frameworks developed in other disciplines, which can be integrated into requirement modelling processes. This in turn enables subsequent evaluation processes [23, 32].

The complexity of the big data analytics presents a formidable challenge for modelling and analysis [11]. Rather than modelling the domain from scratch, we brought cognitive models into the requirement engineering to analyse the features of data and the details of user activities. In this article, underpinned by sense-making models, we proposed a conceptual architecture to understand the user requirements and system characteristics of big data analytics. Specially, we emphasize that a big data analytics solution consists of components that exploit both machine capability and human intelligence. To facilitate and ensure the integration of machine capability and human intelligence, integration-centric components are needed to provide seamless experience of users. The fundamental goal of a solution is to support the big data transformation from raw format to knowledge products.

In summary, this chapter makes the following contributions to the literature.

- A sensemaking perspective to understand big data analytics, which emphasises the human aspects of big data analytics.
- A generic conceptual architecture, which illustrates the essential components and their relationship to provide effective and comprehensive IT support for big data analytics.
- A demonstration of two instantiations of the generic architecture of two use cases to provide examples of big data solutions relative to a situation in a specific organization.

This approach opens up an extra channel to requirements modelling and analysis, which is based on transforming and analysing theoretical models from social science and cognitive science to a design artefact. The research work reported in this chapter provides an illustration of how theoretical models were selected and applied to the analysis and design of the architecture. We hope this modest attempt at bringing social science or cognitive science models into requirement engineering will complement the traditional requirement modelling process. Much more work is needed to refine our method to meet the practical needs of requirements analyst and engineers.

References

1. Agrawal, D., et al.: Challenges and opportunities with big data. Proc. VLDB Endow. **5**(12), 2032–2033 (2012)
2. Barton, D., Court, D.: Making advanced analytics work for you. Harv Bus. Rev. **90**(10), 78–83 (2012)
3. de la Calle, G, Alonso-Martínez, E., Tzagarakis, M., Karacapilidis, N.: The dicode workbench: a flexible framework for the integration of information and web services. In: Proceedings of the 14th International Conference on Information Integration and Web-based Applications and Services (iiWAS2012), Bali, Indonesia, 3–5 Dec 2012, pp. 15–25 (2012)
4. Dervin, B.: From the mind's eye of the user: the sense-making qualitative-quantitative methodology. In: Dervin, B., Foreman-Wernet, L., Lauterbach, E. (eds.) Sense-Making Methodology Reader: Selected Writings of Brenda Dervin. Hampton Press Inc, Cresskill (2003)
5. Fisher, D., DeLine, R., Czerwinski, M., Drucker, S.: Interactions with big data analytics. Interactions **19**(3), 50–59 (2012)
6. Gartner, Inc.: Pattern-based strategy: getting value from big data. Gartner Group press release, July 2011. http://www.gartner.com/it/page.jsp?id=1731916
7. Henschen, D.: Why all the hadoopla? Inf. Week **11**(14), 11 (2011)

8. Hevner, A.R., March, S.T., Park, J., Ram, S.: Design science in information systems research. MIS Q. **28**(1), 75–105 (2004)
9. Karacapilidis, N.: Mastering data-intensive collaboration and decision making through a cloud infrastructure: the dicode EU project. EMBnet. J. **17**(1), 3 (2011)
10. Karacapilidis, N., Rüping, S., Tzagarakis, M., Poigné, A., Christodoulou, S.: Building on the synergy of machine and human reasoning to tackle data-intensive collaboration and decision making. Intelligent Decision Technologies. Springer, Berlin, 2011, pp. 113–122
11. Kiron, D., Shockley, R., Kruschwitz, N., Finch, G., Haydock, M. Analytics: the widening divide. MIT Sloan Manage. Rev. **53**(3), 1–20 (2011)
12. Kittur, A., Chau, D.H., Faloutsos, C., Hong, J.I.: Supporting Ad hoc sensemaking: integrating cognitive, hci, and data mining approaches. In: Sensemaking Workshop at CHI, Boston, MA (2009)
13. Klein, G., Moon, B., Hoffman, R.R.: Making sense of sensemaking 2: a macrocognitive model. Intel. Syst. IEEE **21**(5), 88–92 (2006)
14. Lee, C.P., Abrams, S.: Group sensemaking. In: Position Paper for Workshop on Sensemaking. ACM Conference on Human Factors and Usability (CHI), Florence, Italy (2008)
15. Lim, E.-P., Chen, H., Chen, G.: Business intelligence and analytics: research directions. ACM Trans. Manage. Inf. Syst. (TMIS) **3**(4), 17 (2013)
16. Lohr, S.: The age of big data. New York times, Feb 11, 2012 (2012). http://www.nytimes.com/2012/02/12/sunday-review/big-datas-impact-in-the-world.html
17. Maiden, N., Jones, S. Karlsen, K., Neill, R., Zachos, K., Milne, A.: Requirement engineering as creative problem solving: a research agenda for idea finding. In: Proceedings of 18th IEEE International Conference on Requirements Engineering (RE'10). IEEE Press, pp. 57–66 (2010)
18. March, S.T., Smith, G.F.: Design and natural science research on information technology. Decis. Support Syst. **15**(4), 251–266 (1995)
19. Ntuen, C.A., Balogun, O., Boyle, E., Turner, A.: Supporting command and control training functions in the emergency management domain using cognitive systems engineering. Ergonomics **49**(12–13), 1415–1436 (2006)
20. Paul, S.A., Reddy, M.C.: Understanding together: sensemaking in collaborative information seeking. In: Proceedings of the 2010 ACM Conference on Computer Supported Cooperative Work. ACM (2010)
21. Peffers, K., Tuunanen, T., Rothenberegr, M.A., Chatterjee, S.: A design science research methodology for information systems research. J. Manage. Inf. Syst. **24**(3), 45–77 (2007)
22. Pirolli, P., Card, S.: The sensemaking process and leverage points for analyst technology as identified through cognitive task analysis. In: Proceedings of International Conference on Intelligence Analysis, vol. 5 (2005)
23. Qu, Y., Furnas, G.: Model-driven formative evaluation of exploratory search: a study under a sensemaking framework. Inf. Process. Manage. **44**(2), 534–555 (2008)
24. Qu, Y., Hansen, D.L.: Building shared understanding in collaborative sensemaking. In: Proceedings of CHI 2008 Sensemaking Workshop, 2008
25. Russell, D.M., Stefik, M.J., Pirolli, P., Card, S.K.: The cost structure of sensemaking. In: Proceedings of SIGCHI. ACM Press, New York, pp. 269–276 (1993)
26. Savolainen, R.: The sense-making theory: reviewing the interests of a user-centered approach to information seeking and use. Inf. Process. Manage. **29**(1), 13–18 (1993)
27. Schoenfeld, A.H.: Learning to think mathematically: problem solving, metacognition, and sensemaking in mathematics, In: Grouws, D. (ed.) Handbook of Research on Mathematics Teaching and Learning. MacMillan, New York (1992)
28. Thakker, D., Dimitrova, V., Lau, L., Denaux, R., Karanasios S., Yang-Turner, F.: A priori ontology modularisation in ill-defined domains. In; Proceedings of the 7th International Conference on Semantic Systems. ACM (2011)
29. The Economist: Data, data everywhere. Feb 25, 2010. http://www.economist.com/node/15557443

30. Weick, K.E.: Sensemaking in Organizations. Sage Publications Inc, Thousand Oaks (1995)
31. Yang-Turner, F., Lau, L.: A pragmatic strategy for creative requirements elicitation: from current work practice to future work practice. In: Workshop on Requirements Engineering for Systems, Services and Systems-of-Systems (RESS), 2011. IEEE (2011)
32. Yang-Turner, F., Lau, L., Dimitrova, V.: A model-driven prototype evaluation to elicit requirements for a sensemaking support tool. In: Proceedings of the 2012 19th Asia-Pacific Software Engineering Conference, vol. 1. IEEE Computer Society (2012). doi: 10.1109/APSEC.2012.129
33. Yu, E.: Modelling strategic relationships for process reengineering. In: Yu, E., Giorgini, P., Maiden, N., Myopoulous, J. (eds.) Social Modeling for Requirements Engineering. The MIT Press, Cambridge (2011)
34. Zeng, L., Li, L., Duan, L.: Business intelligence in enterprise computing environment. Inf. Technol. Manage. **13**(4), 297–310 (2012)

Chapter 4
Making Sense of Linked Data: A Semantic Exploration Approach

Dhavalkumar Thakker, Vania Dimitrova, Lydia Lau,
Fan Yang-Turner and Dimoklis Despotakis

Abstract There are growing arguments that Linked Data technologies can be utilised to enable user-oriented exploratory search systems for the future Internet. Recently, search over Linked Data has been studied in different domains and contexts. However, there is still limited insight into how conventional semantic browsers over Linked Data can be extended to empower exploratory search, which is open-ended, multi-faceted and iterative in nature. Empirical user studies in representative domains can identify problems and elicit requirements for innovative functionality to assist user exploration. This chapter presents such an approach—a user study with a unifocal semantic data browser over several datasets linked via domain ontologies is used to inform what intelligent features are needed in order to assist exploratory search through Linked Data. We report main problems experienced by users while conducting exploratory search tasks, based on which requirements for algorithmic support to address the observed issues are elicited. A semantic signposting approach for extending a semantic data browser is proposed as a way to address the derived requirements.

Keywords Semantic data exploration · User interaction · Linked data · Exploratory search · Empirical requirements elicitation

D. Thakker (✉) · V. Dimitrova · L. Lau · F. Yang-Turner · D. Despotakis
University of Leeds, Leeds LS2 9JT, UK
e-mail: D.Thakker@leeds.ac.uk

V. Dimitrova
e-mail: V.G.Dimitrova@leeds.ac.uk

L. Lau
e-mail: L.M.S.Lau@leeds.ac.uk

F. Yang-Turner
e-mail: F.Yang-Turner@leeds.ac.uk

D. Despotakis
e-mail: scdd@leeds.ac.uk

N. Karacapilidis (ed.), *Mastering Data-Intensive Collaboration and Decision Making*,
Studies in Big Data 5, DOI: 10.1007/978-3-319-02612-1_4,
© Springer International Publishing Switzerland 2014

4.1 Introduction

Linked Data technologies have received wider acceptance, both in industry and academia [1]. One of the major factors for this success has been the availability of large amount of semantic data in various formats and domains (http://lod-cloud. net/state/). In parallel with engineering solutions for seamless generation of semantic data, efforts have been made to facilitate user interaction with such data. There are growing arguments that Linked Data technologies can be utilised to enable user-oriented exploratory search systems for the future Internet [2]. In contrast to regular search, exploratory search is open-ended, multi-faceted, and iterative in nature, and is commonly used in scientific discovery, learning, and sense making [3, 4].

There are a wide range of tools available for offering exploratory search using semantic web technologies (state-of-the-art in [5, 6]). However, exploratory search over linked data is still insufficiently studied. As pointed in a recent keynote focusing on interaction with Linked Data [7], although the technological platforms for exploring linked data are growing, enabling citizen users to explore interconnectable links associated with structured data is still a key challenge. This calls for anurgent attention by researchers and technology developers to identify major issues with user exploration of linked data, derive requirements for new methods, and engineer solutions to implement these methods utilising semantic technologies and tools. Experimental studies with existing systems in domains well-presented in linked data can be used to elicit requirements for engineering new methods for user exploration [8].

The work presented in this chapter follows the above arguments, and specifically focuses on providing intelligent functionality embedded in a data browser to assist users in their exploratory search tasks over linked data. This is part of an ongoing research examining intelligent interfaces for interactive sensemaking over Linked Data, conducted intheframework of the EU project Dicode (http:// dicode-project.eu) which develops intelligent services for data intensive decision making and collaboration. We have built a fairly traditional semantic data browser—Pinta—which provides a base line for identifying key issues users face with conventional uni-focal exploratory search interfaces over linked data. An instantiation of Pinta in the Music domain—MusicPinta—is used in an experimental study with users to elicit requirements for intelligent assistance based on observations of challenges users face while interacting with MusicPinta; and suggesting a way to address them by adding signposting features. The work contributes to the engineering of intelligent web applications over Linked Data by providing key requirements, and an approach to address them, elicited with an empirical requirements elicitation method and applicable to exploratory search over linked semantic data.

The chapter will position the research within the relevant literature pointing main contribution (Sect. 4.2). Section 4.3 will present the base line system, including both the generic architecture (Pinta) and its instantiation in the music domain (MusicPinta), which provides a testbed for eliciting requirements and designing new functionality. A user study with MusicPinta is presented in Sect. 4.4, briefly outlining key findings. Section 4.5 reports observations of main interaction issues faced by users, based on which requirements for adding intelligent functionality are elicited. Following the requirements, a signposting approach for adding intelligent assistance is proposed (Sect. 4.6). The chapter concludes by pointing at future work.

4.2 Related Work

Tools for exploratory search using linked data. One class of applications developed to facilitate exploratory search focuses on faceted search using linked data. Faceted search works by suggesting restrictions as facets, i.e. selectors for subsets of the current set of items [5]. The work presented in this chapter examines the role of semantic tags and their effect while browsing and learning in another class of applications called semantic data browsers. Such browsers operate on semantically augmented data (e.g. semantically tagged content) and layout browsing trajectories using relationships in the underpinning ontologies. Tabulator [9] can be considered as the first semantic data browser which enables users to browse data by following semantic links to resources. Two types of semantic data browsing have been emerged since—(i) pivoting (or set-oriented browsing) and (ii) multi-pivoting. In a pivoting browser, a many-to-many graph browsing technique is used to help a user navigate from a set of instances in the graph through common links [6]. Exploration is often restricted to a single start point and uses 'a resource at a time' to navigate in a dataset [10]. This form of browsing is also referred as uni-focal browsing. The second type of browsers—multi-pivoting—allows a user to start from multiple points of interest, i.e. multi-focal exploration of maps, by zooming multiple parts of the map at the same time [11]. The state-of-the-art of semantic data browsers is covered in [6]. In this chapter, we present a fairly traditional semantic data browser, Pinta, which provides a uni-focal interface for browsing through several linked semantic datasets related to music domain. Our contribution to semantic data browsers is identifying key challenges for exploratory search users face, and deriving requirements on how to extend a browser to address these challenges.
Empirical requirements elicitation for exploratory search using linked data. Recent work examines search over linked data. Research and evaluations of alternative approaches (than keyword search) to data exploration for knowledge building are seen as preparation for the next generation of Web, the Web of Linked Data [8]. This brings forth for the need of empirical requirements elicitation studies, as the one presented here. In the five year's reflection on evaluating semantic search systems [12], key requirements are identified from the perspective of performance evaluation of semantic search systems. We present requirements elicited by involving users in interaction with the system, which complements the

requirements in [12]. Notably, the user study steps on conducting user studies for deriving requirements for exploratory search interfaces. Following recommendations in [13], we examine the cognitive load when conducting exploratory search tasks (in our case, this is done utilising a modified version of NASA-TLX questionnaire [14]).

There have been workshops on the topic of challenges of user interactions, e.g. Semantic Web User Interactions (SWUI), Intelligent Exploration of Semantic Data (IESD) and Exploratory Search series. These events have produced useful guidelines and shared experiences about outstanding challenges, requirements and methodologies to follow, e.g. [15]. Our work contributes to these ongoing efforts by focusing specifically on the effect of the browser on users' ability to complete exploratory search tasks and identifying requirements for further intelligent support to facilitate fruitful exploration, suggesting also a signposting approach.

4.3 Baseline System for Browsing Through Linked Semantic Data

In this section, we present a traditional semantic data browser called Pinta,[1] which provides a uni-focal interface for browsing through several linked semantic datasets. While Pinta is generic, its instantiation in a specific domain (Music in this case) is prepared to provide a platform for empirical requirements elicitation.

4.3.1 Pinta: A Generic Uni-focal Semantic Browser Shell

The main goal of Pinta is to enable users to easily tap into resources built from the Web and, in particular, exploring the use of the Linked Data paradigm. Figure 4.1 depicts the traditional three-layer architecture for Pinta which comprises: (i) the Data Layer, including knowledge sources and content, (ii) the Processing Layer, including modules for semantic augmentation and query, and (iii) the Presentation Layer for content browsing. The implementation of Pinta combines state-of-the-art semantic web technologies for semantic augmentation, semantic query and data representation.

The **Data Layer** contains domain specific ontological knowledge sources and content assembled from the Web (Linked Data and other, domain specific, sources). The knowledge sources consist of graphs of ontological concepts relevant to the domain of interest. They provide the foundation for semantic augmentation of the content in the Processing Layer, and the structure for semantic trajectories for

---

[1] Using an analogy with Christopher Columbus' ship 'La Pinta'; in our case, Pinta is a browser shell providing a means to explore through a vast amount of data.

Fig. 4.1 Architecture of the generic uni-focal semantic data browser Pinta

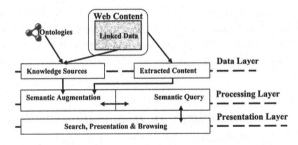

Fig. 4.2 A faceted-layout template for presenting a focus (currently explored) entity in Pinta

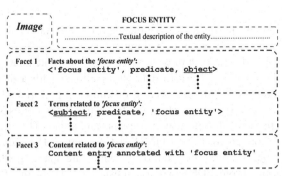

browsing in the Presentation Layer. The content is in textual format and can be assembled from more than one online platform, e.g. blogs, reviews, comments.

The **Processing Layer** has two main services: (i) semantic augmentation of the assembled content and (ii) semantic queries to retrieve content for the Presentation Layer. Semantic augmentation (also known as semantic tagging) is a process of attaching semantics (in the form of ontology concepts) to a selected part of text. The semantic augmentation module in Pinta includes: Semantic Repository, Information Extractor and Semantic Indexer. The Semantic Repository (using OWLIM) combines the functionality of an RDF-based DBMS and an inference engine. The Information Extractor (using GATE—General Architecture for Text Engineering) produces annotated sets of extracted entities with offset, ontology URI and type information. The Semantic Indexer (using Sesame SPARQL API) converts these annotated sets to RDF triples. Semantic Query service takes term(s) or concept(s) as keywords and output information relating to the matching concept(s) and content(s). Together with the Semantic Repository, Semantic Queries implement various concept/content lookup functionalities to find related and relevant concept(s) or content(s) from the Semantic Repository.

The **Presentation Layer** provides a front-end for the output of semantic queries from the Processing Layer (the template for a focus entity is shown in Fig. 4.2). The interface layout includes three main facets and a description (at the top) extracted from the knowledge datasets for the focus entity (being currently explored): (i) Facet 1 includes facts about the focus entity; (ii) Facet 2 includes terms related to the focus entity; and (iii) Facet 3 shows related content.

Facts and related terms for the focus entity consist of triples from the Semantic Reposity, which include hierarchy links (denoted as is a kind of), membership (denoted as is a type of) and object properties (denoted as other). Hyperlinks are provided to further details for the retrieved objects.

4.3.2 MusicPinta: An Instantiation of Pinta in the Music Domain

The Music domain has been selected for an instantiation of Pinta and has been used as a testbed to observe exploratory search and derive requirements for intelligent support. The Web of data is rich in music-related datasets and content. As of 2011, there were at least 13 datasets identified, with a diverse range of concepts and ambiguous entities covering instruments, performances/events, artists, and music genres. The data sets used for MusicPinta comprise the following resources. DBpedia[2]: for musical instruments and artists. This dataset is extracted from dbpedia.org/sparql using CONSTRUCT queries. These queries along with a programming wrapper and additonal coding are made available as open source at the sourceforge.[3] DBTune[4]: for music-related structured data made available by the DBTune.org in linked data fashion. Among the datasets on DBTune.org we utilise: (i) Jamendo—a large repository of Creative Commons licensed music; (ii) Megatune—an independent music label; and (iii) MusicBrainz—a community-maintained open source encyclopedia of music information . Amazon reviews for musical instruments shown in Pinta. All datasets, except the reviews, were available as RDF datasets and the Music ontology[5] was used as schema to interlink them. The Amazon reviews were converted in RDF using Pinta's semantic augmentation of textual content in the Processing Layer.

The datasets provide an adequate setup (fairly large and diverse data set, yet of manageable size for experimentation) for examining user behaviour during exploratory search. It has 2.4 M entities and 38 M triple statements, taking 1.5 GB physical space and includes 876 musical instruments ontology entities, 71 k performances (albums, records, tracks) and 188 k artists. The datasets coming from DBTune.org (such as MusicBrainz, Jamendo and Megatunes) already contain the "sameAs" links between them for linking same entities. We utilise the "sameAs" links provided by DBpedia to link MusicBrainz and DBpedia datasets. This way, the DBpedia is linked to the rest of the datasets from DBtune.org, thus enabling exploration via the rich interconnected datasets.

---

[2] http://dbpedia.org/About

[3] http://sourceforge.net/p/pinta/code/38/tree/

[4] http://dbtune.org/

[5] http://musicontology.com/

4.4 User Study and Interaction with MusicPinta

To observe user exploratory search behaviour and elicit requirements for adding intelligent functionality in uni-focal semantic browsers over linked data, we conducted an exploratory study with MusicPinta.

4.4.1 Study Design

The study involved 12 participants recruited on voluntary basis (a compensation of £15 Amazon vouchers was paid). Half of the participants were native speakers and the other half spoke and communicated in English fluently. All participants had IT background, good experience in web search. Each participant attended an individual session, conducted and observed by an experimenter for an hour. This session comprised the following steps:

- using a Pre-study questionnaire [5 min] for collecting information about the user and test his/her domain awareness;
- introducing MusicPinta [10 min];
- conducting Task 1 [15 min] aiming at identifying distinctive characteristics of the musical instrument "bouzouki";
- conducting Task 2 [15 min] for identifying usage and features of the musical instrument "electric guitar";
- a Post-study questionnaire [10 min] for testing again the participant's domain awareness and gathering usability feedback; and,
- briefly interviewing [5 min] foreliciting the overall impression of using MusicPinta for exploratory search.

After each task (third and fourth step), the users were asked to fill-out a short questionnaire to assess their cognitive load using a modified version of the NASA-TLX questionnaire [14]. The data collected in the study includes: (i) the forms with completion of Tasks 1 and 2; (ii) the pre- and post-study questionnaires, and (iii) the system log data.

4.4.2 User Interaction with MusicPinta

The study required participants to complete two tasks related to exploring musical instruments and was positioned within an advertising scenario for a fictitious UK music shop (see Table 4.1). In both tasks, the participants were given an entry point for browsing and asked to fill in their answers in a provided template. The tasks exhibit the characteristics of exploratory search tasks summarised in [16]: the main goal is learning and/or investigation of a musical instrument; there is a low

Table 4.1 User tasks in the experimental study

Task 1: Characteristics of a musical instrument (bouzouki)	Task 2: Usage and features of a musical instrument (electrical guitar)
The music shop is extending its collection of instruments with international musical instruments. You work in an advertising agency which has been asked to prepare an advertisement script for some of the new instruments that will appear in the shop. A key part of the preparation of the advertisement script is the research of the product	The music shop wants to increase the sales of its traditional musical instruments, such as electrical guitars. It intends to do this by adding links to creative commons album recordings with electric guitars, together with some interesting information about these albums to inspire customers to play/buy electric guitars or other musical instruments
You have been asked to conduct a research of one of the new instruments, called bouzouki, using the information available in MusicPinta. You have to identify: • Other main characteristics of bouzouki; • up to five similar instruments to bouzouki; • features that make bouzouki distinctive from the similar ones you have chosen	Furthermore, when displaying its electric guitar items, the shop wants to highlight key features people look for when purchasing electric guitars You are asked is to conduct the research to address the above requirements by using information provided in MusicPinta. You have to review the information about electric guitar and identify: • three interesting album recordings that include electric guitars and specify what is interesting; • key features that people look for when purchasing an electric guitar
Go to 'Semantic Search' in MusicPinta and type bouzouki. Browse the content and follow links. Complete the provided form	Go to 'Semantic Search' in MusicPinta and type electric guitar. Browse the content and follow links. Complete the provided form

level of specificity about the information needed and how to find it; search is open ended, requires finding several items and involves a degree of uncertainty; tasks are 'not too easy' and include multiple facets.

The completion of Task 1 required mainly browsing through the musical instrument classification (in both DBTune and DBpedia) and reading descriptions provided from DBpedia. The task was analytical in nature, as users had to perform comparison and identification of distinctive features. Example screen shots from a user's interaction are shown in Fig. 4.3 (the user reviews the information provided on Bouzouki—the image, description, categorisation, and other terms), and Figs. 4.4 and 4.5 (examining the content of similar instruments selected from the facet Plucked string instrument available on the Bouzouki interface).

In contrast, Task 2, which required browsing through content about music albums (see Fig. 4.6) and artists, and reading through Amazon reviews, was more ambiguous and involved some creative thinking and imagination.

Home

Semantic Search

Contribute

Bouzouki

The bouzouki is a musical instrument with Greek origin in the lute family. A mainstay of modern Greek music, the front of the body is flat and is usually heavily inlaid with mother-of-pearl. The instrument is played with a plectrum and has a sharp metallic sound, reminiscent of a mandolin but pitched lower. There are two main types of bouzouki. The three-course has three pairs of strings (known as courses), and the four-course has four pairs of strings.

Facts about "Bouzouki"   [-] Collapse

Bouzouki	is a kind of	Instrument
		Plucked string instruments
		String instruments
Bouzouki	is type of	Instrument
Bouzouki	subject	Cypriot musical instruments
	subject	Greek musical instruments
	subject	Necked bowl lutes
	subject	Turkish loanwords

Terms related to "Bouzouki"   [-] Collapse

Lute	is a kind of	Bouzouki
Sitar		Bouzouki
Xalam (khalam)		Bouzouki
Oud		Bouzouki
Cittern		Bouzouki
Rebab		Bouzouki
Morten Musicus performing (recorded on album M□n...	instrument	Bouzouki
Lanvall performing (recorded on album The Grand De...	instrument	Bouzouki
Warren Ellis performing (recorded on album The Lyr...	instrument	Bouzouki
Morten Musicus performing (recorded on album M□n...	factor	Bouzouki

Fig. 4.3 "Bouzouki"—facts, related terms, description, image, tagged content (obscured)

Lute

Home

Semantic Search

Contribute

Lute can refer generally to any plucked string instrument with a neck (either fretted or unfretted) and a deep round back, or more specifically to an instrument from the family of European lutes. The European lute and the modern Near-Eastern oud both descend from a common ancestor via diverging evolutionary paths.

Facts about "Lute"   [-] Collapse

Lute	is a kind of	Bouzouki
		Bouzouki
		Instrument
		Lute
		Lute
		Plucked string instruments
		String instruments
Lute	is type of	Instrument
Lute	subject	Arabic words and phrases
	subject	Baroque instruments
	subject	Composite chordophones

Fig. 4.4 "Lute"—facts, related terms (obscured), description, image, tagged content (obscured)

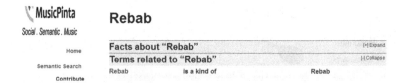

Fig. 4.5 Information on "Rebab". Unlike Lute or Bouzouki, semantic datasets have no significant information on Rebab (e.g. image, description, facts and related terms were either missing or did not contain new information)

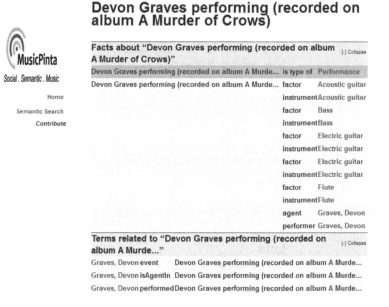

Fig. 4.6 A performance involving Electrical Guitar. The page is typical of performances. Facts and terms were generally present but textual description or media was generally not available

4.4.3 Data Analysis and Results Outline

The detailed analysis on the task performance and learning outcome in the user study is covered in a recent publication of our group [17]. In this section, we have only focused on the important aspects of the task outcome that allow us to comment during our analysis of the browsing behaviour.

Two musical instrument experts (one for Bouzouki, one for Electric guitar) have marked the outcome of participants for the two tasks. The marking is to measure how successful the participants have been in completing the tasks using MusicPinta. For Task 1, all together, the participants identified 44 characteristics (70 %, individual score median 4) from the description section of bouzouki (including the picture), and 19 descriptions (30 %, individual score median 1.5) from the semantic tags. Task 2 was also completed reasonably well (average score 48 %).

Table 4.2 Abstraction level assigned to the user semantic facet clicks

Abstraction level	Description
Classification—upper level	Clicks on abstract entities, such as instrument, performance, artist, from the Music Ontology
Classification—middle level	Clicks on middle level entities such as classification of musical instruments, e.g. string instruments, plucked string instruments, steel guitars, Greek musical instruments
Classification—low level	Clicks on entities at concrete/concrete level, e.g. representing musical instruments instances, e.g. bouzouki, mandolin, lute, electric guitar
Content—albums	Clicks on descriptions of music albums
Content—artists	Clicks on descriptions of music artists
Content—reviews	Clicks on Amazon reviews of musical instruments

In terms of participants' subjective perception of both tasks, while the average performance on Task 2 was significantly lower than Task 1 (Wilcoxon test, $W = 74$, $p < 0.005$), there was no significant difference in the participants' confidence scores. Task 2 was more frustrating with border line significance (Wilcoxon test, $W = -34$, $p < 0.05$). There was no correlation between scores and confidence—participants seemed confident that they did as much as they could in the given time.

The interaction log files, which recorded the user clicks when browsing datasets in MusicPinta, provided an insight into the browsing behaviour. MusicPinta recorded every user click with following information: user id, timestamp and ontology entity of the link. The ontology entity of a link consists of the source of dataset (e.g. dbpedia or dbtune) and the entity (e.g. Ukulele or plucked string instrument). The log data was pre-processed and each link was assigned an abstraction level based on the ontology (Table 4.2). The distribution of user clicks on both tasks according to the abstraction levels is presented in Fig. 4.7.

In both tasks, the participants were asked to start browsing from an entity classified at low level, i.e. a concrete musical instrument (bouzouki and electric guitar, respectively). Participants clicked much more on entities from the middle level (instrument classifications) in Task 1 than in Task 2. Similarly, participants clicked much more on content entities in Task 2 than in Task 1. The observations of key exploratory search challenges users faced with MusicPinta informed requirements for further extension and are listed next.

4.5 Requirements for Assisting User Browsing Over Linked Data

The following observations are based on the study considering the user interaction while browsing the linked semantic data. Each observation was assessed to elicit requirements for supporting exploratory environments.

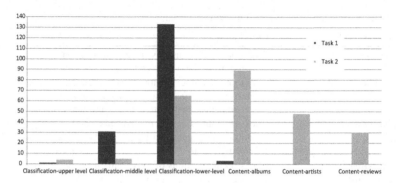

Fig. 4.7 Distribution of user clicks in both tasks (total no. of clicks in each abstraction level)

Observation 1: Abstraction conundrum. While browsing specific instruments (e.g. Bouzouki in Fig. 4.3 or Lute in Fig. 4.4), performances (e.g. Devon Graves performance in Fig. 4.6) and performers, two participants clicked on abstract concepts, such as instrument, performance and performer, from the Music Ontology. In both cases, the participants were looking for concrete information (e.g. participant-12 clicked on instrument in Task 1 when seeking for more detail about a musical instrument, while participant-05 clicked on performer and performance in Task 2 when seeking more detail about an album). The aggregated datasets in MusicPintahave large number of instances for the abstract concepts (which is typical of linked datasets, see Sect. 4.3.2), which led to confusion as the result was a long list of performers, performance and instruments, and the participants quickly pressed the back button on their browsers.

Requirement 1: Offering semantic links at an appropriate level of abstraction. The above observation motivates consideration on identifying what can be algorithmically offered as the right level of abstraction on various browsing junctures. This is especially important when the abstract concepts have large amount of concrete instantiations. The main challenge here is what to suppress and what to display to the user; e.g. how to decide which performances out of 71 k to display when a user is on the entity page of the abstract concept performance.

Observation 2: Exploring entities/content with insufficient information. Another interesting case is the high number of 'empty clicks'—the user clicks on a link and is taken to a page with no information, sees that this link is not helpful and quickly returns to the previous page. In Task 1, such clicks concerned similar instruments, e.g. there was no information about bajitar, xalam, rebab (see Fig. 4.5). In Task 2 such clicks concerned performances (music albums) and happened quite often. 'Empty clicks' leading to pages with no information was seen as one of the main reasons foruser's frustration. At the same time, may be due to their experience of links that lead to dead ends, some links were perceived as empty without exploring them further and the users missed to click on important for the tasks information (e.g. pages about musical instruments were abandoned, although there was useful information about relevant instruments; or interesting facts about an album artist were overlooked as the users did not click on the corresponding link). With linked datasets, it is typical to find entities that do not have much explanation or links to other entities.

Similar issues were observed with content (Amazon Reviews in our user study). Textual contentin semantic data browsers are semantically tagged and made available via one of the facets (see facet 3 in Fig. 4.2). Users clicked to view some of the Amazon reviews to find out more information about an instrument and its review. However some of the reviews were deemed to have insufficient information to be useful. This observation is in line with relevant research conducted which concludes that not all reviews are equally helpful (for example, [18] identifies useful reviews have considerable review depth compared to non-useful reviews). One can extrapolate such observations to be generic enough to be applied to social content and conclude that social content has avariety of usefulness levels, while being possible to find content that has insufficient information to be of help in browsing.

Requirement 2: *Reduce entity link options*. Avoid showing entity links that do not lead to any new information. Reduce number of entity links shown to the user based on their browsing value; allowing reduction of clutter and confusion. The challenge here is to define what 'browsing value' is and how to calculate it for an entity with respect to other entities from the same entity page.

Requirement 3: Reduce content link options. Avoid showing content links that do not lead to any new information. Reduce number of content links shown to the user based on their helpfulness/usefulness. The challenge here is to decide the parameters of helpfulness/usefulness of content.

Observation 3. Varied selection strategies while facing too many choices. Both tasks (deliberately) put the users in situations where they had too many choices. This meansthat the users had a large number of links to review while in a focus entity page. For example, the bouzouki page included 12 different links in the facts facet (11 links to concepts in the middle classification level and 1 link to the abstract concept instrument) and 51 links in the terms facet (43 links to musical instruments and 8 links to performances). The entry point in Task 2, the electric guitar page, included 18 links in the facts facet (to concepts in the middle and upper classification levels), 78 links to albums in the terms facet, and 8 links to Amazon reviews in the content facet. This is a typical situation with the datasets from linked data. For example, for the DBpedia dataset, which has 3.5 M entities and 627 M triples, on average, auser might have to review 192 links while exploring a focus entity.

We observed users following different strategies when presented with too many choices in the browsing interface: (i) clicking on the nearest classification link from the 'facts facet' (e.g. plucked string instruments or string instruments) to see general characteristics in the case of bouzoukias part of Task 1. However, users rarely clicked on links from the facts facet as part of the Task 2, as the task did not require this; (ii) clicking on instruments mentioned in the 'related terms facet'— (e.g. lute and mandolinmainly in Task 2; (iii) clicking on something (e.g. 'an instrument') that 'sounds familiar' (e.g. sitar, banjo, pipain Task 1); (iv) click on something (e.g. 'an instrument or an album') that sounds interesting or unusual (e.g. oud, xalamin Task 1 and noticing a women artist or something interesting in the album name in Task 2); (v) clicking on something that looks important (e.g. an artist has several albums in Task 2); and, (vi) clicking randomly (after exhausting other strategies).

This observation is in line with the latest research in search engines and HCI; increasing numbers of options can make designers and users feel less confident when deciding and less happy with theresults [19, 20]. To support the varied level of selection strategies, two requirements are derived.

Requirement 4. Make facet selection processdynamic and intuitive. The use of different types of facets is useful but dependent on tasks. The challenge here is to give control to users and help them to decide or make it easier for them to decide the facet and when they would like to use it. For example, providing guideline on the utility of different types of facets in the system and allowing facets to be added or removed as required by the user.

Requirement 5. Take into account context to cater for interests and importance. People when faced with many choices do select what they find useful/ familiar/interesting/unusual/important. Hence, there is a merit in making it easier for users to decide/spot easily these values? The challenge here is how to measure and decide these values from the available options for a specific user or holistically.

Observation 4. Text and Media information influence user experience and performance. For Task 1, a great deal of performance owed to the use of textual description and images while identifying characteristics of an instrument. Hence, there is value in offering unstructured (textual and multimedia information) in conjunction with structured data (semantics) for exploration.

Requirement 6. Offer relevant multimedia or textual information. The exploration tools developers shall carefully select multimedia or textual information for the domain and make them available as part of the focus entity pages. For example, in MusicPinta instruments and performances related pages can contain YouTube videos of instruments or performances involving instruments.

4.6 Semantic Signposting to Assist Exploratory Search

The identified requirements from the study indicated the need for further algorithmic support to realise the exploratory search potential of semantic data browsers. There can be many possible ways to address these requirements. One of such possible approach is *semantic signposting* which was implemented as part of the Dicode project and presented here.

In uni-focal exploration, a user focuses on one entity at a time represented in a page. This entity page contains links to various descriptions, image and also links to other entities. Such entity page can be treated as a juncture in the journey where the explorer has to make few choices (through the links which takes him to different paths). Some of the requirements elicited (R1–R3, R5) can be addressed by providing different types of signposts guiding the explorer in making a choice about paths she can take: *Only showing 'important' links*, which are a subset of all possible links for the user to review as part of next path he/she can take.

Let us call *'candidate entities'* all the links possible to navigate from a focus entity page. Then, the importance of each candidate entitycan be computed based on density parameters such as—number of further entities available from a candidate entity (i.e. number of directly connected entities to the candidate entity), number of potentially reachable entities from this entity (i.e. number of indirectly connected entities or number of entities connected to candidate entity via directly connected entities) and type and weight of the connection (e.g. semantic relationship between the candidate entities and it's directly connected entities). The judgement of creating subset of links can be implemented using density metrics for the semantic graphs [21], where density function shall allow comparing how dense/informative each of the path is originating from a candidate entity. The subset of links to be shown to the user in this case will be based on the density value of each link (i.e., candidate entity).

Showing indicator of importance for each link. For more creative tasks (such as Task 2 in our study), which require browsing through a large amount of content, the study appeared to provide indication that it will *not be very beneficial to limit the user entity choices*, as this can affect the free content exploration. Instead, signposting can include some *indicators about the 'importance' or 'value' of a content item*, e.g. if there is any description (or any multimedia content), its source of the content (e.g. DBpedia, MusicBrainz), if further semantic links are available in the content (e.g. albums that have several musical instruments) to facilitate user choices. There can be some ordering based on the value. Again the judgement of importance can be implemented using density metrics for the semantic graphs as discussed in the previous point.

Adaptive signposts. One of the other parameters to consider while judging importance of links is consideration of user's prior knowledge, e.g. does user already know about a particular entity or class of entity? Such consideration in creation of signposts (i.e. reducing number of links shown to the user) can allow users to decide what is useful/familiar/interesting/unusual/important (R5). However, detecting prior knowledge is generally challenging, especially when the freedom of exploration has to be preserved. A possible way to 'sense' previous knowledge is to analyse the user clicks on the low classification level links—clicking on an instrument can indicate some familiarity with its most specific classification category (e.g. in the study, users familiar with Russian musical instruments clicked on Balalaika and users familiar with Chinese musical instruments clicked on Pipa). The necessary techniques to address such requirements can benefit from the research in the user modelling, adaptation and personalisation. Such solution can allow creating signposts that include familiar and new knowledge together. Putting familiar and new items together in such a way can deepen the learning by association [22].

4.7 Conclusions

We have presented a study with a traditional uni-focal semantic data browser to observe browsing behaviour of users while interacting with several linked semantic datasets aiming at deriving requirements to inject intelligent features. We have found several intricate challenges that are applicable to typical interaction over linked semantic datasets. For example, disparity of the options available while browsing from an entity. In some cases large number of links available from an entity, hence posing too many options for the user to choose from and in other cases no links or information available making users frustrated. We have also observed and reported varied levels of selection strategies when a user is faced with too many options.

These requirements are useful for the researchers and developers in the area of user interactions over linked data. There can be multiple approaches to address such requirements. One of such approaches is presented in this chapter with the concept of *semantic signposting*. The signposting can allow guiding users to subsets of 'important' links for browsing. The 'importance' of links can be judged on the basis of amount of possible navigation paths or steps from the entity in focus. We also for see the benefits of considering user's prior knowledge to adapt signposts in the semantic data browsers.

As a continuation of our work, we are implementing signposting functionality in Pinta. With this extension, we will conduct a comparative user study with the current system presented in this chapter as a baseline. We also intend to involve a large number of participants to exploit appropriate quantitative analysis techniques.

Acknowledgments The authors thank the participants in the study.

References

1. Bizer, C., Heath, T., Berners-Lee, T.: Linked data—the story so far. Int. J. Semant. Web Inf. Syst. **5**(3), 1–22 (2009)
2. Waitelonis, J., Knuth, M., Wolf, L., Hercher, J., Sack, H.: The path is the destination–enabling a new search paradigm with linked data. In: Proccedings of the Workshop on Linked Data in the Future Internet at the Future Internet Assembly, Ghent, Belgium, CEUR Workshop Proc., 16–17 Dec (2010)
3. Marchionini, G.: Exploratory search: from finding to understanding. Commun. ACM **49**(4), 41–46 (2006)
4. White, R.W., Kules, B., Drucker, S.M., Schraefel, M.C.: Supporting exploratory search, introduction, special issue, communications of the ACM. Commun. ACM **49**(4), 36–39 (2006)
5. Hermann, A.: Semantic search: reconciling expressive querying and exploratory search. In: Aroyo, L., Welty, C. (eds.) The Semantic Web—ISWC 2011, pp. 177–192. Springer, Heidelberg (2011)

6. Popov, I., Schraefel, M., Hall, W., Shadbolt, N.: Connecting the dots: a multi-pivot approach to data exploration. In: The Semantic Web ISWC 2011 10th International Semantic Web Conference, Bonn, Germany, 23–27 Oct 2011
7. Schraefel, M.C.: What does it look like, really? Imagining how citizens might effectively, usefully and easily find, explore, query and re-present open/linked data. In: Proceedings of the 9th International Semantic Web Conference on the Semantic Web, vol. 6497, pp. 356–369 (2010)
8. Wilson, M.: From keyword search to exploration: designing future search interfaces for the web. Found. Trends® Web Sci. 2(1), 1–97 (2011)
9. Berners-lee, T., Chen, Y., Chilton, L., Connolly, D., Dhanaraj, R., Hollenbach, J., Lerer, A., Sheets, D.: Tabulator: exploring and analyzing linked data on the semantic web. Methodology 2006(1), 6 (2006)
10. Schwabe, D.: Explorator: a tool for exploring RDF data through direct manipulation. Linked data on the web WWW2009 workshop (LDOW2009) in conjuction with WWW 2009, vol. 538. CEUR Workshop Proceedings, Madrid, Spain (2009)
11. Javed, W., Ghani, S., Elmqvist, N., Lafayette, W.: PolyZoom: multiscale and multifocus exploration in 2D visual spaces. In: Proceedings of the 2012 ACM Annual, pp. 287–296 (2012)
12. Uren, V., Sabou, M., Motta, E., Fernandez, M., Lopez, V., Lei, Y.: Reflections on five years of evaluating semantic search systems. Int. J. Metadata Semant. Ontol. 5(2), 87–98 (2010)
13. Chandler, P., Sweller, J.: Cognitive load theory and the format of instruction. Cogn. Instr. 8(4), 293–332 (1991)
14. Hart, S.G., Staveland, L.E.: Development of NASA-TLX (task load index): results of empirical and theoretical research. Hum. Ment. Workload 1(11), 139–183 (1988)
15. White, R.W., Muresan, G., Marchionini, G.: Report on ACM SIGIR 2006 workshop on evaluating exploratory search systems. ACM SIGIR Forum 40(2), 52 (2006)
16. Wildemuth, B.M., Freund, L.: Assigning search tasks designed to elicit exploratory search behaviors. In: Proceedings of the Symposium on Human–Computer Interaction and Information Retrieval—HCIR 12, pp. 1–10 (2012)
17. Dimitrova, V., Lau, L., Thakker, D., Yang-turner, F., Despotakis, D.: Exploring exploratory search: a user study with linked semantic data. In: ACM Workshop on Intelligent Exsploration of Semantic Data (IESD2013), Conjunction with Hypertext 2013, pp. 9–16 (2013)
18. Mudambi, S.M., Schuff, D.: What makes a helpful online review? a study of customer reviews on Amazon.com. MIS Q. 34(1), 185–200 (2010)
19. Oulasvirta, A., Hukkinen, J., Schwartz, B.: When more is less: the paradox of choice in search engine use. In: Proceedings of the 32nd international ACM SIGIR conference on Research and development in information retrieval (SIGIR '09), pp. 516–523. ACM, New York, NY, USA (2009)
20. Schwartz, B.: The Paradox of Choice: Why More Is Less, 1st edn, p. xi, 265. ECCO, New York (2004)
21. Alani, H., Brewster, C.: Ontology ranking based on the analysis of concept structures. In: Proceedings of the 3rd International Conference on Knowledge Capture KCAP 05, p. 51 (2005)
22. Roschelle, J.: Learning in interactive environments: prior knowledge and new experience. In: Falk, J.H., Dierking, L.D. (eds.) Public Institutions for Personal Learning: Establishing a Research Agenda, p. 37–51. American Association of Museums, Washington (1995)

Chapter 5
The Dicode Data Mining Services

Natalja Friesen, Max Jakob, Jörg Kindermann, Doris Maassen, Axel Poigné, Stefan Rüping and Daniel Trabold

Abstract Real world problems in society, science or economics need human structuring, interpretation and decision making, the limiting factor being the amount of time and effort that the user can invest in the sense-making process. The Dicode data mining services intend to help in clearly defined steps of the sense-making process, where human capacity is most limited and the impact of automatic solutions is most profound. This includes recommendation services to search and filter information, text mining services to search for new information und unknown relations in data, and subgroup discovery services to find and evaluate hypotheses on data. This chapter provides an overview of the data mining services developed in the context of the Dicode project. It addresses the usability of the services and indicates which big data technologies are being used to deal with very large data collections.

N. Friesen (✉) · J. Kindermann · A. Poigné · S. Rüping · D. Trabold
Fraunhofer IAIS, Schloss Birlinghoven, 53754 Sankt Augustin, Germany
e-mail: friesen.natalja@googlemail.com

J. Kindermann
e-mail: joerg.kindermann@iais.fraunhofer.de

A. Poigné
e-mail: axel.poigne@iais.fraunhofer.de

S. Rüping
e-mail: stefan.rueping@iais.fraunhofer.de

D. Trabold
e-mail: daniel.trabold@iais.fraunhofer.de

M. Jakob · D. Maassen
Neofonie GMBH, 10115 Berlin, Germany
e-mail: jakob@neofonie.de

D. Maassen
e-mail: doris@neofonie.de

N. Karacapilidis (ed.), *Mastering Data-Intensive Collaboration and Decision Making*,
Studies in Big Data 5, DOI: 10.1007/978-3-319-02612-1_5,
© Springer International Publishing Switzerland 2014

Keywords Data mining · Text mining · Data-intensiveness · Big data · Services · Usability

5.1 Introduction

The amount of data generated in the world is growing at an unprecedented rate. Key drivers are novel technologies (such as next-generation sequencing or new imaging technologies in medicine), the trend towards ubiquitous data generation and recording (ranging from smart phones over business process automatisation to integration of IT and production technology in Industry 4.0), and new ways of data production and sharing driven by revolutions in society such as the mass use of social media and crowdsourcing. A key challenge for future business, research and society is to make sense of this growing amount of data and enable humans to understand it.

The term "big data" was coined to describe this situation of massive production of novel types of data and the technologies that facilitate it. In one of the most widely accepted definitions, big data is described by three aspects:

- **Volume**: the size of individual data sets is increasing, and more and more often breaks the barrier where it can be readily stored and processed by single machines. Computing clusters are the standard hardware solution, driving the need for distributed data processing solutions.
- **Velocity**: data is frequently generated at high speeds, e.g. by automated business process or technical systems, and requires analyses on-the-fly with little delay to automatize reactions.
- **Variety**: data is not clearly structured, as for example in a relational database, but comes in a variety of forms. For example, the mixture of classical structured data and text data is very prominent in medical sciences and in social media applications.

Novel technologies are being developed to deal with these types of data. However, while it is clear that a huge amount of data can be stored and processed using available technologies, the question of the value of storing all this data is still significantly less clear. The basic problem is the following: in order to generate value out of data, it is necessary to analyze, dissect and understand the data, and extract valuable new knowledge; this can be significantly harder in the situation of big data. The curse of dimensionality, spurious correlations and problems of significance of multiple hypothesis testing are well-understood problems that make automatic analysis of highly variable real-world data hard from the perspective of statistics. Simply put, from the data analytics perspective, the key question is: *if data analytics can be compared to finding the needle in the haystack, does big data analytics mean more needles or just more hay?*

The driving motivation for data mining in the context of Dicode was that in order to help people collaborate and make better decisions based on available data, it is strictly necessary that the data mining technologies should facilitate a better interpretation of the data and a better understanding of the knowledge that is hidden inside data. Real world problems in society, science or economics cannot be solved by machines alone, but need human structuring, interpretation, and decision making. What can be observed is that the limiting factor in big data analytics is the amount of time and effort that the user can invest in the sense-making process. The Dicode project has hence investigated data mining approaches that help in clearly defined steps of the sense-making process, where human capacities is most limited and the impact of automatic solutions is most profound. This includes recommendation services to search and filter information, text mining services to search for new information und unknown relations in data, and subgroup discovery services to find and evaluate hypotheses on data.

In particular, text mining technologies (such as named entity recognition, named entity disambiguation, relation extraction, and opinion mining) have reached a level in which it is—for the first time—practically feasible to apply semantic technologies to very large data collections. Dicode's approach in text mining makes use of the huge advancement in the development of scalable data mining frameworks and technologies (most of them exploiting the cloud computing paradigm), such as MapReduce, Hadoop, and Mahout.

Dicode's text mining services aim at supporting different types of users and facilitating text analysis within the use cases of the project. The services can be easily targeted to specific tasks, such as analysis of social media. They can help organizations to quickly spot trends, allowing them to become proactive, rather than reactive, to market conditions. The text mining services make sense of background knowledge, while properly leveraging Linked Data, Big Data and data visualization.

5.2 An Overview of the Dicode Data Mining System

The Dicode Data Mining framework consists of a set of REST-based services. The Dicode Data Mining services aim to support different types of users and facilitate data analysis. Through their integration into the Dicode Workbench (see Chap. 7), they enable users to benefit from the Dicode's decision support mechanisms. The Dicode Data Mining services can be classified in the following groups:

Group 1: General Data Mining services. These provide a declared functionality and can be easily targeted to specific tasks. They include:

- Subgroup Discovery
- Recommender Service
- Similarity Learning Service
- RapidMiner Service
- Embedded R Executor.

Group 2: Instances of particular Data Mining services. Services of Group 1 are adapted to a specific task so that they can be directly applied by users without any particular data mining knowledge. This group consists of the following services:

- Subgroup Discovery for Genomic Data Analysis
- Recommender of GEO Datasets.

Group 3: Text Mining services. These services enable the processing and analysis of unstructured data, in particular text. The full list of these services is as follows:

- Twitter Harvester
- Twitter Pre-processing Service
- Blog Pre-processing Service
- Named Entity Service
- Entity Prominence Service
- Topic Detection Service
- Phrase Extraction Service
- Phrase Extraction training Service
- Emotion Detection Service
- DBpedia Spotlight Named Entity Service
- Log Aggregation Service
- Opinion Mining Service.

5.3 General Data Mining services

5.3.1 Subgroup Discovery

Subgroup discovery is a knowledge discovery task that aims at finding subgroups of a population with high generality and distributional unusualness with respect to the target attribute. A subgroup is a set of terms $\{t_1, \ldots, t_k\}$ where every term t_i is a constraint on an attribute, i.e. t_i has the form $(a_i = v_i)$, v_i in $D(a_i)$, where $D(a_i)$ is a domain of the attribute a_i. The length of the subgroup description is the number of terms it is built of.

The *Subgroup Discovery (SD) service* uses a specialized in-memory database using specialized data structures like *FP-trees* [1]. The majority of subgroup discovery algorithms typically rely on top–down search combined with considerable pruning, which exploits anti-monotonicity of the quality measure. When dealing with high-dimensional data, the hypothesis space becomes extremely large, and the whole discovery process becomes overly time-consuming. The SD algorithm used in Dicode's SD service uses weighted covering strategy [2], which was proven to perform well in comparison with other similar algorithms [3].

SD is a method that is often used to generate a human understandable representation of the most interesting dependencies in the data. Hence, the more crisp and concise the output is, the better. Unfortunately, standard algorithms often produce very large and redundant outputs. It is hard for a researcher to make use of the results, particularly for large and complex data. In order to reduce the output space, we extended the SD service by a component for finding relevant subgroups. The newly developed approach is described in [4]. This modification leads to a considerable reduction in the amount of returned patterns without loss of their statistical descriptiveness, and as a consequence, to a better understanding of the data. Another important issue concerning interpretability of the results is to enable a user to influence the output by including/excluding certain attributes from the search. Receiving a feedback on their results, the user can set up a new iteration of the algorithm by specifying undesired (e.g. biological or medical) attributes. We extended the Subgroup Discovery Service by a component that attends this task.

To sum up, we offer a SD service as a standard method for data analysis. We developed a new mechanism for finding relevant subgroups, which is important to get non-redundant results. We extended the service in order to be able to include/exclude particular terms from the search. Our SD service is generic enough to be applied in a wide range of research fields. Despite its generality, the service can be easily targeted to more specific tasks, such as analysis of genomic data.

5.3.2 Recommender Service

In the fields of medical and biomedical research (see Chap. 8), tools are needed to reduce the information overload, provide advice in finding an interesting dataset or medical report, and facilitate decision-making. Recommendation exploits user feedback to predict the "preference" that a user would give to an item not seen before. The *Recommender Service* and *Similarity Learning Service* address this issue. The services form two necessary components of a recommendation mechanism. While the Recommender Service provides the user with relevant and interesting information, Similarity Learning Service aims to create a similarity model from the user preferences, which can be used for user specific recommendations. Both services support a general framework that can be easily adapted to recommendation of a wide range of information objects. We illustrate this in the example of Recommender System for Gene Expression Omnibus (GEO) datasets, which was created using both services (see Sect. 5.4.2 of this chapter).

The Recommender Service ranks information items with respect to their importance to the user. In some cases, it is hard to formally model the user's interests. It may be easier for the user to define an object of interest instead. Given a particular information item the user is interested in (the "reference object"), the most similar items will be then recommended to the user. The service first computes a similarity between the item the user is interested in and each item among the candidate items. In the second step, the algorithm ranks all items according to their similarity to the reference object and returns the most similar ones. This method is computationally intensive since its computation time is increasing linearly to the number of comparisons. This computational burden can be reduced by pre-structuring the data, e.g. using *Antipole tree* indexing as proposed in [5]. To reduce the computation time, Dicode adopted this approach and integrated it into the service.

The Dicode Recommender framework, which consists of the Recommender Service and the Similarity Learning Service, was especially developed for recommendation of scientific items. The recommender systems in a scientific context are significantly different from the standard case of product recommendations. The biggest issues are the representation of complex objects, a smaller—more heterogeneous—set of users and a lack of information about the user's preferences. One of the most important requirements for such systems is their ability to provide personalized recommendations. As users working on a collaborative problem come from various disciplines—the typical usage scenario being small teams of people working on different projects—a single recommendation scheme might not cover the needs of all users. Thus, recommendations need to be more personalized.

For ease-of-use purposes, the user should not be burdened with the additional trouble of customizing the recommendation. Instead, the system should use machine learning techniques to adapt itself to the user's preferences. This issue is addressed by the Similarity Learning service, which aims to create a similarity model from the user's feedback. The very important advantage of this approach is that it avoids the so called "cold start problem", which is common for systems based on the collaborative filtering approach. Collaborating filtering is based on collecting users' profiles. A typical profile consists of aggregated information about a user's preferences that are represented by a set of rated items. The user is recommended items that people with similar taste liked. A more detailed comparison of recommender systems in e-science with common approaches can be found in [6]. The Dicode Recommender framework avoids the "cold start problem" by an appropriately designed sampling strategy and provides the user with personalized recommendations.

5.3.3 Similarity Learning Service

In principle, the Similarity Learning Service is a generic learning framework that can operate on a wide range of information items from different research fields. Given an information object, the service delivers a similarity model that is learned from user feedback—a set of object pairs labeled as "similar" or "dissimilar". Therefore, the most important prerequisites for model creation are the availability of training data and a set of basic similarities that are defined by a domain expert. The service is specific enough to learn an accurate similarity model that delivers good results.

Due to the variety of information objects, the question of how to define a similarity for them is a challenging task. It is even more complicated in case the definition of similarity depends on user needs. It makes no sense to generalize a recommendation function among multiple users. There is a requirement for an easy-to-use service that each user can create a similarity model according to his/her particular preferences.

At the same time, a typical user will be unwilling to spend a lot of time to set up a recommendation system. The process of obtaining labeled data is costly in terms of time and manual effort. In order to start learning with n examples, the user needs to give his feedback for $n*(n-1)$ object pairs. Hence, he should be only asked for input that he can give quickly and correctly. In particular, it is very favorable to ask the user only questions regarding specific instances, for which domain experts can usually give very concrete feedback. As an example, when recommending papers to read, it is better to ask the user "is this paper relevant to you?" instead of "do you like to see more papers of the same author?"

In order to reduce the user's efforts in labeling, we investigated how to select a small set of pairs that is informative enough to create an accurate model. We developed an intelligent sampling strategy that selects the most 'interesting' pairs from a pool of unlabeled data to show them to the user.

We exploited the idea of describing massive data sets by using latent components proposed in [7]. Specifically, we adopted the algorithm presented in [7] for sampling the most 'informative' instance pairs. The main idea is to describe data by representing it as a linear combination of dominant latent components. Let $V = [v_1, ..., v_n]$ be a data matrix. It can be decomposed into a product of two lower rank matrices W and H: $V \approx WH$. The matrix W determines a basis of extreme points $W = [w_1, ..., w_k]$, $k \ll n$. $H = [h_1, ..., h_n]$ contains the mixing coefficients that result from solving constrained quadratic minimizing programs. This yields basic vectors that usually correspond to the most extreme data points. Moreover, these vectors span over a simplex that encloses most of the remaining data. To find the most 'informative' instance pairs, we set a data matrix $V = P$ (a matrix of instance pairs represented by numerical vectors).

The described approach was integrated in the Similarity Learning Service. This extension contributed significantly to the quality and usability of the service, since the new sampling strategy enables to create an accurate personalized similarity model on the cost of minimal manual efforts from the user.

5.3.4 RapidMiner Service

RapidMiner is a widely used open source data mining workbench suitable for many data mining tasks. It features a graphical interface to design custom analysis workflows, which enables non data mining experts to easily design their first data mining workflow. RapidMiner comes with a wide set of functionalities for:

- Process control
- Import
- Export
- Data generation
- Data Transformation
- Modeling
- Evaluation.

A wide set of well-known machine learning techniques from the following areas are available and were tested:

- Classification and Regression
- Clustering
- Association Rule and Item Set Mining
- Correlation and Dependency computation.

These include:

- Support Vector Machines
- Artificial Neural Networks
- Decision Trees
- Naive Bayes Classifiers
- Logisitic Regression
- FP-Growth
- Several discretization strategies
- And many other.

The wide set of algorithms makes it easy to try different algorithms on a data set to find a suitable model. These models may be as simple as decision trees, which are easy to understand, or as complex as artificial neural networks.

The Dicode *RapidMiner Service* offers the functionality of RapidMiner within Dicode as a REST service. In particular, it features a workflow and result repository to share workflows and results with other users. This contributes towards the goal of collaborative data analysis and decision making. The benefit of this service grows with the number of shared workflows. However, a desktop installation of Rapid Miner is required to edit workflows.

5.3.5 Embedded R Executor

Clinico-genomic research prescribes some specific requirements to the analysis tools. Many good software modules for statistical analysis of genomic data are offered as open source. One of the most important platforms for these is *R* (free, open source). We offer the Embedded R Executor Service as a second data mining platform.

5.4 Particular Data Mining Services

5.4.1 Subgroup Discovery for Genomic Data Analysis

Applied to gene expression data, the standard SD service would deliver a set of gene names that share similar properties relative to the research question of interest, i.e. SD uses the filtered dataset as produced by the statistical methodology

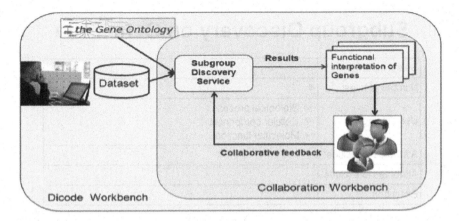

Fig. 5.1 Interactive SD service for functional interpretation of genomic data

used (see Chap. 8). The translation of these results into useful biological knowledge still remains a necessary validation procedure, which is often time-consuming. For instance, one might wonder how the set of genes can be described in terms of molecular or cellular function. Knowledge databases such as *Gene Ontology* (GO, http://www.geneontology.org/) or *Kyoto Encyclopedia of Genes and Genomes* (KEGG, http://www.genome.jp/kegg/) serve as an excellent basis for the interpretation of genes. The integration of these additional knowledge databases into the Subgroup Discovery Service would provide the researcher with more meaningful results.

Another important issue is to enable the user to share their results with other researchers that are working on similar problems and to discuss weaknesses of the identified patterns. Taking these issues into account, the enhanced version of SD service includes the following extensions:

- Integration of external knowledge databases, such as Gene Ontology
- Adaptation of the SD Service to the task of functional interpretation of gene sets
- Development of a user friendly interface
- Integration into the Dicode's collaboration workspaces (see Chap. 6).

Figure 5.1 illustrates the interactive character of the SD service. First, the user specifies parameters such as a category of GO and the number of retrieved subgroups (Fig. 5.2). Then, the service identifies the most interesting subgroups and displays them in the collaboration workspace (Fig. 5.3). Each retrieved subgroup is presented as a single element (XML document) and can be discussed separately.

In summary, we provide a service for functional interpretation of genomic data. The service is based on a subgroup discovery algorithm and includes external knowledge databases (GO). The service has an interface that enables the user to customize the output.

To deal with Genomic Data Analysis, the Subgroup Discovery Service has been extended by:

Subgroup Discovery on Gene Data

Select input file:		Browse...
Number of rules:	5	
Use ontology:	☑ Biological process ☑ Cellular component ☑ Molecular function	
Attributes to include:		
Attributes to exclude:		
Execution:	Go!	

≫dicode

Fig. 5.2 Interface of SD service

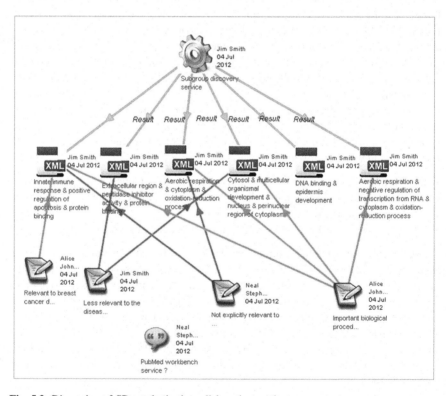

Fig. 5.3 Discussion of SD results in the collaboration workspace

- **Integration of external knowledge databases**. Gene Ontology (GO) serves as a controlled vocabulary of terms for describing genes according to several aspects. GO includes three ontologies containing the description of molecular functions, biological processes and cellular locations of any gene product, respectively. Within each of these ontologies, the terms are organized in a hierarchical way, according to parent–child relationships in a directed acyclic graph (DAG). This allows a progressive functional description, matching the current level of experimental characterization of the corresponding gene product. Currently, the three components of the GO are integrated in the service. Moreover, we aim to include further knowledge databases in the future, such as the KEGG or Reactom (http://www.reactome.org/ReactomeGWT/entrypoint. html).

- **Adaptation of the SD Service for the functional interpretation of gene sets**. The main purpose of a typical microarray experiment is to find a molecular explanation for a given macroscopic observation. The most common methods are based on a 'functional enrichment'. First, genes of interest (e.g. genes that are significantly over- or under expressed when two classes of experiments are compared) are selected. Then, external sources of information, such as gene ontologies and pathways databases, are included to translate the set of genes into interpretable biological knowledge. We extend the SD service by a component that transforms the dataset submitted by the user into a large list of genes enriched by GO terms. We adopted an approach first presented in [8].

5.4.2 Recommender of GEO Datasets

The analysis of in-house data is often restricted by the small size of datasets. However, an enrichment of data by additional datasets can significantly improve the results. As a high number of datasets can be generated and stored relatively easily, it becomes increasingly hard to keep an overview of them. Dataset repositories such as *Gene Expression Omnibus* (GEO) tackle this problem by offering the possibility to publish, explore and query archived datasets. An even higher amount of support is given by different tools integrated into a repository. However, for a single scientist it is difficult to stay aware of all the work being done that may be relevant to him. In order to find an interesting dataset, the user has to define the search criteria. The search for an appropriate dataset is a time-consuming manual process. Due to the complexity of research issues, the user would benefit from recommendations that were generated automatically according to his interests. There is clearly a need for a recommender system that facilitates the reuse and retrieval of datasets.

Taking into account this issue, we developed the Recommender and Similarity Learning Services to provide the user with relevant and interesting datasets from GEO repository. GEO is the largest public repository for high-throughput gene expression data. It was initially set up to store gene expression data generated by

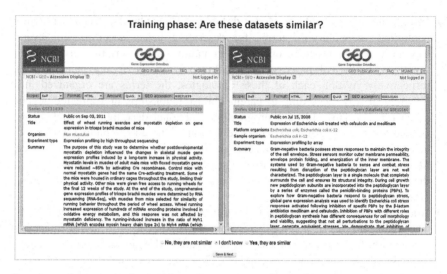

Fig. 5.4 Similarity learning service: interface to obtain a user feedback about the similarity of a dataset pair

microarrays and serial analysis of gene expression. To facilitate the usage of the services, we developed a user friendly interface.

A dataset in the context of the GEO repository is an item that defines a set of related 'Samples' considered to be part of a study and describes the overall study aim and design. A 'Sample' record is composed of a description of the biological material, the experimental protocols to which it was subjected, and a data table containing abundance measurements for each feature. Each stored dataset is created by a specific user, is associated with a set of samples, and contains text information such as title, description and summary. The similarity model was set according to an expert feedback about the importance of single attributes. We have set in the service the following list of basic attribute-similarity measure pairs, which compose the similarity model used for recommendation of GEO datasets:

- Q-Gram distance on title text
- Q-Gram distance on summary text
- Cosine distance on overall study design text
- Cosine distance on experiment type
- Dice similarity on organism
- Euclidean similarity on the number of samples.

The importance of each single attribute-measure pair is obtained from user feedback in the similarity learning process. The user is shown pairs of datasets and is asked to mark them as "similar"/"dissimilar"/"don't know". The presented pairs are selected according to the sampling algorithm, described in Sect. 5.3.3. Figure 5.4 shows an example of a dataset pair presented to the user in the learning process.

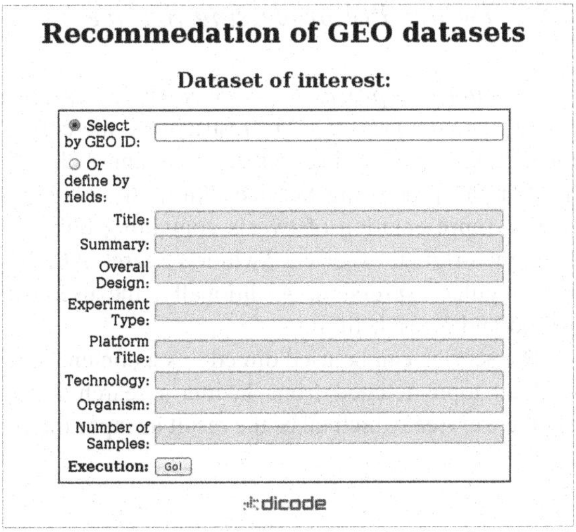

Fig. 5.5 Interface of
Recommender Service

Once a similarity model is learned, the user can start a recommendation process. Figure 5.5 presents the GUI interface for the recommendation service. The user can formulate his needs in two different ways: by defining a dataset of interest from GEO repository or by filling in some text field. The service returns a list of the k datasets that best satisfy the criteria defined by the user.

5.5 Text Mining Services

5.5.1 Twitter Harvester

Twitter Harvester is Dicode's tweet acquisition component. Tweets are restricted to those available via Twitter's Streaming API, which returns about 1 % of the tweets published worldwide. A project in need of a significant amount of tweets will have to use one of Twitter's commercial data providers. Twitter partners with data providers like *Gnip* and *Datasift* (https://dev.twitter.com/docs/twitter-data-providers), which offer the complete Twitter stream and provide custom filters and Twitter pre-processing. We expect that in the future Social Media Analysts in need of Twitter analytics will receive their analysis either directly from Twitter or from specialized platforms that provide analysis of the complete Twitter stream. For Dicode's use cases, data sources like blogs, forums and business news are generally much more attractive than Twitter data. As there is no need for a configuration component anymore, the Twitter Harvester does not provide a service interface but serves as the data storage layer that is accessed via the HBase API directly.

5.5.2 Twitter Pre-processing Service

The *Twitter Pre-processing Service* serves as an interface to Dicode's Twitter corpus. Due to Twitter's API restrictions, the corpus contains only recent Twitter data. Applications using this service should therefore fetch data regularly. The Twitter Pre-processing Service returns the corresponding Tweet ids and a condensed representation of tweets containing only significant nouns. The user can fetch the original tweet via Twitter's Search API. The query can be either a string or a regular expression. Additionally, the user can specify a language, a time period and a result limit.

The service can be used directly as a content service, which displays tweets for a given topic fetched from Twitter's search API. It can also be chained with another service, which takes the result as input for arbitrary text mining operations.

5.5.3 Blog Pre-processing Service

The *Blog Pre-processing Service* acts as an interface to Dicode's weblog corpus, which contains weblog entries from English and German weblog of various domains. The API differs from the Twitter pre-processing API in just one point: besides significant nouns, the full text of a blog post is also returned.

5.5.4 Named Entity Service

Named Entity Recognition (NER) is the task of the identification and classification of proper names in natural-language text. Dicode's *Named Entity Service* identifies named entities of the following types: PERSON, PLACE, ORGANISATION and WORK. The context dependency is a big challenge in Named Entity Recognition: for instance, the entity "Germany" can be a location in a geographical context or an organization in political context. Named entity disambiguation is therefore performed based on the context of the analyzed surface form. The quality of disambiguation usually increases with text size.

The *Named Entity Service* uses a combination of different techniques that have been proven to provide good results. *Conditional Random Fields* (CRF) are used for the improvement of precision. Disambiguation is performed based on external knowledge. The use of Wikipedia data for this purpose has been widely discussed in the literature. The Named Entity Service has been developed based on link statistics extracted from Wikipedia and Neofonie's *Alexandria Ontology* that uses *Freebase* (http://www.freebase.com). We use a Hadoop-based Open Source component for Wikipedia extraction (called *pignlproc*) that was extended to meet Dicode's requirements. Wikipedia statistics used for disambiguation include the

Fig 5.6 Named Entity List

probability of a link from a certain surface form to a Wikipedia article (e.g. how often "George Bush" either links to the father or the son), how often a certain term in Wikipedia is showing up with a link or without a link, and how many links are referring to a Wikipedia article.[1] The output of the Named Entity service can be used by the front-end developer directly, e.g. to visualize the document content by showing a list of named entities as a tag cloud. The service can also be used as input for higher-level services, e.g. for topic detection.

Figure 5.6 shows an example visualization of a Named Entity list, which contains all Named Entities found in a news article. The Named Entity service is available for both English and German. The visualization is currently only available in German. It lists Persons ("Personen"), Places ("Orte") and Organisations ("Organisationen") separately. In the example, several persons and one organisation have been recognized.

5.5.5 Entity Prominence Service

The *Entity Prominence Service* returns statistics about the occurrence of Named Entities in news and blog documents within a certain time period (hour, day, week, month, year). The user can either query for a certain entity or retrieve the top entities for a time period. The user can filter by language, domain and entity type. The service builds on Dicode's Named Entity Service.

Based on the service, prominence charts for certain entities can be produced. A chart of widgets might allow for brand comparison, source filtering and language filtering. Figure 5.7 shows an example of such visualizations.

---

[1] Developers interested in using the statistics might have a look at Max Jacob's talk at the Berlin Buzzwords Conference 2012 which explains the extraction of Wikipedia statistics in detail: http://vimeo.com/45123391.

Fig. 5.7 Visualization for Entity Prominence Service

Lists of top entities for a time unit (like "entity of the day") are another option. A widget might for example compare the prominence of political parties in different news sources. The top entities can also be used for the development of higher-level services like event detection. For event detection, "unusual" top entities have to be detected. This could be done by monitoring the baseline of Named Entities which occur in news or social media documents. If an entity becomes popular, which did not show up before or which showed up only rarely in the past, this could indicate an event.

Results of the *Topic Detection Service* are displayed as a topic map. Figure 5.8 shows a topic map that was created from a collection of tweets (http://www.sananalytics.com/lab/twitter-sentiment/). The numbered blue nodes in the graph represent the different topics that have been extracted. The nodes that are labeled by words represent those terms that are most important for a given topic. The network structure of the graph emerges, because some of the defining words are related to several topic nodes (words that are related to one topic only are colored in yellow, those that are related to several topics are colored in orange or red, depending on the number of topic relations.). The visualization provides a quick overview of the topics that are present in a text collection as well as their interrelations.

Users are also able to zoom in on a graph detail, as shown in Fig. 5.9. In this figure, the neighborhood of the term "ballmer" (i.e. the Microsoft's CEO Steve Ballmer) is shown. It is related to the topics 13, 31, and 46.

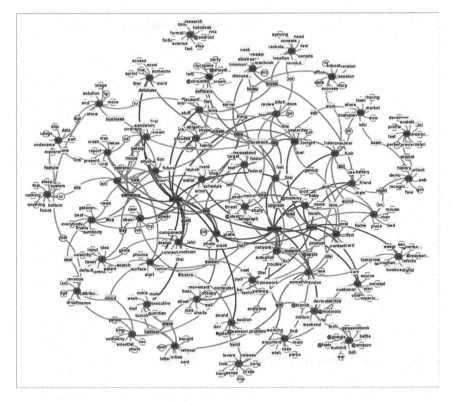

Fig. 5.8 Topic map of a tweet collection

Fig. 5.9 Topic map detail

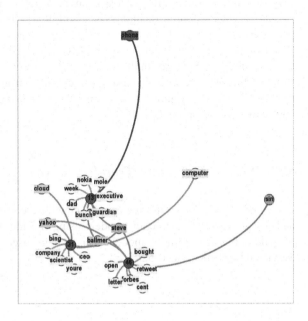

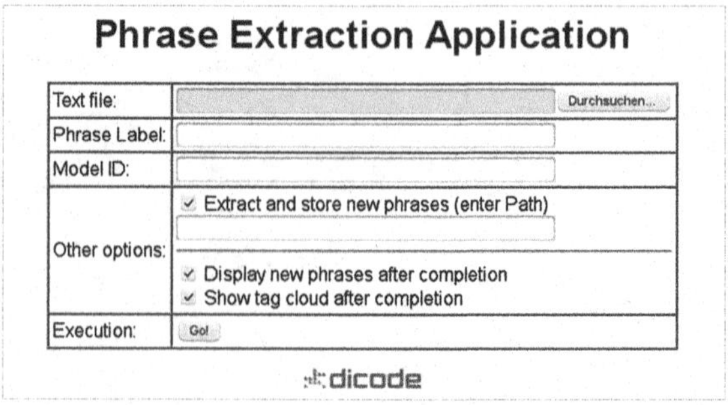

Fig. 5.10 The interface of
the phrase extraction service

Phrase Extraction Application

Text file:		Durchsuchen...
Phrase Label:		
Model ID:		
Other options:	☑ Extract and store new phrases (enter Path)	
	☑ Display new phrases after completion	
	☑ Show tag cloud after completion	
Execution:	Go!	

≈dicode

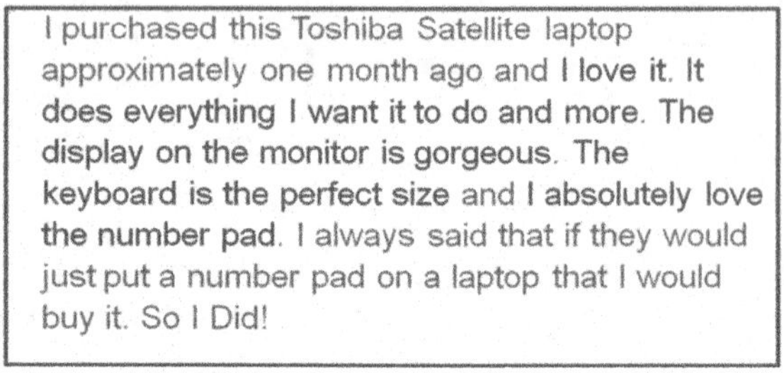

Fig. 5.11 Phrase
highlighting in a text segment

I purchased this Toshiba Satellite laptop
approximately one month ago and I love it. It
does everything I want it to do and more. The
display on the monitor is gorgeous. The
keyboard is the perfect size and I absolutely love
the number pad. I always said that if they would
just put a number pad on a laptop that I would
buy it. So I Did!

5.5.6 Phrase Extraction Service

The *Phrase Extraction Service* allows extracting different types of phrases from a text collection. It is based on the *Conditional Random Field* algorithm. The service uses an active learning strategy that is an extension of the one described in [9].

The service supports an interactive workflow, where the end-user can train phrase extraction models interactively and apply them to a text collection.

The interface of this service (Fig. 5.10) provides two visualizations of the results: highlighted phrases in the texts (Fig. 5.11) and a tag cloud that is generated from the full set of phrases that have been extracted from the whole text collection (Fig. 5.12). These examples have been generated using a collection of Amazon customer reviews on laptops.

5.5.7 Phrase Extraction Training Service

The *Phrase Extraction Training Service* enables the user to interactively develop and test new phrase extraction models using the interface template that is displayed in Fig. 5.13. These models can then be used in the Phrase Extraction Service described in Sect. 5.5.6.

Fig. 5.12 A tag cloud generated from a set of extracted phrases

Fig. 5.13 The interface of the Phrase Extraction Training Service

Phrase Extraction Training

Training text file:		Durchsuchen...
Training phrases file:		Durchsuchen...
Phrase Label:		
Model ID:		
Other options:	☑ Save model locally (enter Path)	
	☑ Extract and store new phrases (enter Path)	
	☑ Display new phrases after completion ☑ Show tag cloud after completion	
Execution:	Go!	

⠐⠅dicode

5.5.8 Emotion Detection Service

The *Emotion Extraction Service* is based on the Phrase Extraction Training service. This service does not need a list of phrases to learn. It only needs a list of wanted seed terms (that may include regular expressions) and a list of terms (or regular expressions) that should not be contained in the phrases. Figures 5.14 and 5.15 provide example expressions. The service will use all phrases around the matching seed terms for the training. Special routines to determine phrase boundaries have been implemented. A list of training phrases is a result of the execution (together with the extraction model). The list may be edited and re-used to train an even better model with the Phrase Extraction Training service.

```
"([Ff]|[Pp]h)antastisch",
"(^|
)([Ww]under|[Aa]ller|[Bb]ild)*schön(e|es|ne|em|er|ere|eres|eren|erem|ste
|stes|ster|sten|stem)($| )",
"(^| )Cool(st)*(e|en|er|em|es)*( |$)",
"(^|
)Schön(e|es|ne|em|er|ere|eres|eren|erem|ste|stes|ster|sten|stem)($| )",
"(^| )Spa(ss|ß)($| )"
"(^| )[Ff]reu(e|st|t|en)( |$)"
"(^| )[Gg]lücklich($| )",
"(^| )[Ww]under(bar|voll)(er)*(e|es|er|en|em)*($| )"
"(^| )gefallen($| )"
"(^| )gefällt($| )"
"(^| )klasse($| )",
"(^| )prima($| )"
"(^| )super($| )"
"(^| )toll(e|es|er|en|em)($| )"
"(ober|super|mords|^| )cool(st)*(e|en|er|em|es)*( |$)",
"[Bb]egeistert"
"[Ee]mpfehlenswert"
"[Gg]eil"
"[Hh]errlich",
"[Hh]ervorragend",
"[Ss]ensationell"
"[Zz]ufrieden",
"hervorragend"
```

Fig. 5.14 Regular expressions of wanted seed terms for the emotion "delight/pleasure" (extract)

The collection of training phrases with the regular expression heuristics simplifies this task considerably and can be performed more effectively than collecting examples manually.

5.5.9 DBpedia Spotlight Named Entity Service

An interactive Named Entity annotation service is required for the following case (which is part of the project's Use Case described in Chap. 9): A marketing professional wants to retrieve additional metadata about all brands found in a single document. For such tasks, we implemented a new service, namely the *DBpedia Spotlight Named Entity Service*.

Dicode's original Named Entity Service does not have the required capabilities, because it annotates named entities in batch mode. We enhanced interactivity by adding an operator to the *MiaQL* processing language, which is available on Neofonie's development cluster. A developer who wants to annotate a newly added document collection can easily annotate named entities in the collection by executing the operator. Despite the improved interactivity, the available Named Entity Service lacks the opportunity to analyze single documents instantly.

The new service is based on DBpedia Spotlight (https://github.com/dbpedia-spotlight), an open source project for automatic annotation of DBpedia (http://dbpedia.org) entities in natural language text. It provides programmatic

```
" (den|einen) [Gg]efallen",
" (zu|mal) schön ",
" [a-zA-Z] klasse",
" gefallen (ist|sind|war|waren)",
" net ",
" net$",
" nie ",
" so schön",
" zwar ",
" ([Aa]uf|[Üü]ber)
(alle|jede[sn]|[Dd]ein(e|en)*|sein(e|en)*|[Ii]hr(e|en)*|[Ee]u(er|re|ren)
|baldige[sn]*|schnelle[sn]*)
)*(Antwort|Meinung|Feedback|Anregung|Input|Foto|Bild|Rück).*      (freue*
ich|freue*n wir)"
" ([Dd]e[nmr]|ganze) [Ss]pa(ß|ss)",
" ([Gg]anz|[Nn]och) schön",
" ([Gg]ar*)*([Nn][ei]t|[Nn]icht*|[Nn]ed)             (mal
)*((absolut|so|sehr|ganz)
)*([Gg]eil|[Cc]ool|[Ee]mpfehlenswert|[Ss]ensationell|[Hh]ervorragend|[Bb
]egeistert|[Zz]ufrieden|[Ss]chön|hervorragend|toll|super|klasse|[Hh]errl
ich|([Ff]|[Pp]h)antastisch|glücklich)",
" ([Hh]ätte|[Kk]önnte|[Ww]ürde)",
" ([Ii]ch     freue*     mich|[Ww]ir     freue*n     uns).*([Aa]uf|[Üü]ber)
(alle|jede[sn]|[Dd]ein(e|en)*|sein(e|en)*|[Ii]hr(e|en)*|[Ee]u(er|re|ren)
|baldige[sn]*|schnelle[sn]*)
)*(Antwort|Meinung|Feedback|Anregung|Input|Foto|Bild|Rück)",
" ([Jj]a|[Nn]a) (prima|super|toll|klasse)",
" ([Jj]a|[Nn]a|[Zz]u) schön",
" ([Mm]edien|[Pp]rofilierungs|[Pp]rofit|[Pp]ublicity)geil",
" ([Nn][ei]t|[Nn]icht*|[Nn]ed)                       (immer
)*([Gg]eil|[Cc]ool|[Ee]mpfehlenswert|[Ss]ensationell|[Hh]ervorragend|[Bb
]egeistert|[Zz]ufrieden|[Ss]chön|hervorragend|toll|super|klasse|[Hh]errl
ich|([Ff]|[Pp]h)antastisch|glücklich)",
" ([Ss]elbst|[Vv]er)herrlich",
" ([Ww]under)*[Ss]chönen guten Tag",
" (^|
) (auf|unter|hinter|über|neben|in) (den|die|das|ein|einen|eine) ([A-Za-z0-
9ßÜüÄäÖö]+ )gefallen",
" (^| ) (nicht*|ne[td]) [Ff]reu(e|t|n|en)*( |$)",
" (ein|aus|runter|raus|auf|rein)gefallen",
" (nur|wenig|bißchen|bisschen|ein|einen|mein|meinen) [Ss]pa(ß|ss)",
" (war|ist) das schön",
"Auge gefallen",
"Dank",
"Entscheidung gefallen",
"Glückwunsch",
"Grüsse",
"Grüße",
"Hände gefallen",
```

Fig. 5.15 Regular expressions of unwanted terms for the emotion "delight/pleasure" (extract)

interfaces for phrase spotting (recognition of phrases to be annotated) and disambiguation (entity linking), as well as various output formats (XML, JSON, RDF etc.) in a REST-based web service. The standard disambiguation algorithm is based upon cosine similarities and a modification of TF-IDF weights (using *Apache Lucene*, http://lucene.apache.org). The main phrase spotting algorithm is *exact string matching*, which uses *LingPipe's Aho-Corasick* implementation (http://en.wikipedia.org/wiki/Aho-Corasick_algorithm).

Fig. 5.16 The interface of the DBpedia Spotlight Named Entity Service

Figure 5.16 shows the user interface of the DBpedia Spotlight Named Entity Service. The user can restrict the entities spotted in the document by selecting types via the "Select types" menu shown in Fig. 5.17.

5.5.10 Log Aggregation Service

During the course of the project, we constantly strived to improve manageability and stability of our distributed computing infrastructure. Debugging distributed systems, like *Hadoop* and especially Hadoop's distributed database *HBase*, can be rather tedious. Tracking down the root cause of an error involves searching multiple gigabytes of log files on various machines. Using standard Unix tools like *grep* (http://en.wikipedia.org/wiki/Grep) for searching is inefficient, because a search on a single machine might already take several minutes. Analyzing log files on a dozen of nodes requires plenty of time. Log aggregation has therefore become standard for the operation of large software systems like Hadoop clusters. Besides log aggregation, commercial products like *Splunk* (http://www.splunk.com) or open source software like *Logstash* (http://logstash.net) offer a convenient search interface. Dicode's *Log Aggregation Service* uses Logstash.

Logstash permits the usage of a variety of technologies for processing and storage of log files and thus offers both the performance and flexibility required to process such potentially high volume data. Our setup is based on *Redis*

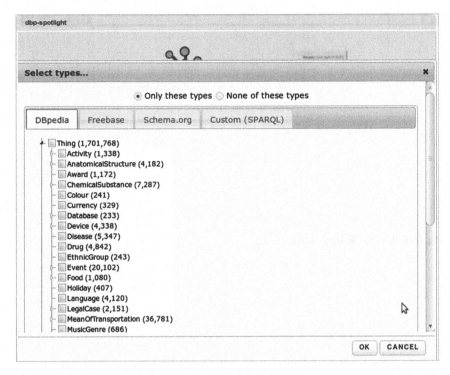

Fig. 5.17 Selecting types (DBpedia Spotlight Named Entity Service)

(http://redis.io), which is used as event queue, and *Elasticsearch* as persistence layer (Elasticsearch is an increasingly popular open source search engine based on Apache Lucene, http://www.elasticsearch.org).

Moreover, Logstash has an active community and develops quite fast. Thanks to its modular concept, it is highly extensible. Numerous input modules, filters and output modules for most of the common open source technologies in this area are already available. As front-end application, we use *Kibana* (http://kibana.org), an open source interface to Logstash and Elasticsearch, which is available under the MIT licence.

Figure 5.18 shows the Logstash interface. Via the timeline, the user can zoom into the log messages. The menu on the left of Fig. 5.19 offers filtering and/or highlighting capabilities. Additionally, Logstash offers a streaming view that serves the incoming messages instantly as shown in Fig. 5.20.

Due to security restrictions, the Log Aggregation Service is only available within Neofonie's internal network. Access from outside the network is not necessary because the log analysis interface is only used by developers responsible for the operation of the development cluster. If external parties develop a module which is deployed via the MIA web application, debugging of Hadoop/HBase lies in the hands of Neofonie's developers.

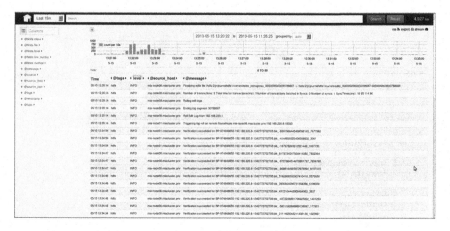

Fig. 5.18 Kibana interface (initial view)

Fig. 5.19 Filtering by host in Logstash Kibana interface

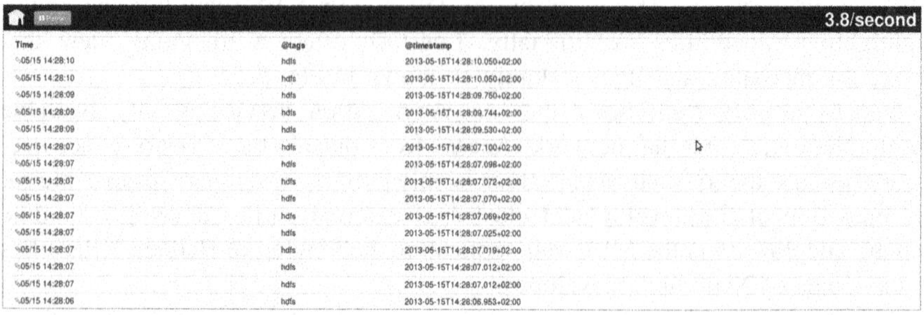

Fig. 5.20 Logstash streaming view

5.6 Using the Services

In this section, we give key examples of the user experience when using the Dicode Data Mining Services. We show that using the corresponding technologies, e.g. topic maps for unstructured data and subgroup discovery for structured data, new insights into data can be gained that are not easily found using manual approaches. In particular, we aim to demonstrate that these data mining results facilitate a high-level discussion of important patterns in the data and thus contribute novel, useful and objective information to the human collaboration and sense-making process. It is noted that the tools developed in Dicode for the end user in most cases use a combination of different services. Typically, a visualization component translates the results into an appropriate visualization.

5.6.1 Topic Detection Service

One of the key challenges that were tackled in the Dicode project consists in summarizing large collections of text data, particularly of social media snippets from blogs and forums, such that a human can understand and single out the most important discussion topics. This approach makes sense because reading and understanding a single text is easy for a user; the challenge comes from finding the right text to read from the vast number of texts that are written daily.

Figure 5.21 shows an excerpt of a Topic Map that has been generated with Dicode's *Topic Detection Service*. Using data from a collection of German car reviews, it shows different topics together with their descriptive keywords and the relations of the topics to each other. For example, "Topic 173" is described by the words "acceleration", "seconds", "speed", "sprint" etc. The analyst can easily detect that this topic describes the driving characteristics of a sportive car, and can use this information to zoom in on all the documents that cover this topic, if required. From the statistics given with the topics (encoded as bubble size), he can also infer that the topic is a fairly frequent one, and how characteristic the keywords are for the topic (encoded as colors).

Furthermore, based on Dicode's *Opinion Mining Service*, topics and keywords can be also correlated with emotions. For a business analyst (e.g. a marketing manager), this offers a new way to navigate through the space of topics and emotions of the customers based on original and objective data.

5.6.2 Subgroup Discovery Service

The *Subgroup Discovery Service* has been integrated into Dicode's Collaboration Support Services (see Chap. 6) in order to integrate automated data analysis in the context of a collaborative discourse. However, the key property that influences the user experience with respect to subgroup discovery is not the implemented user

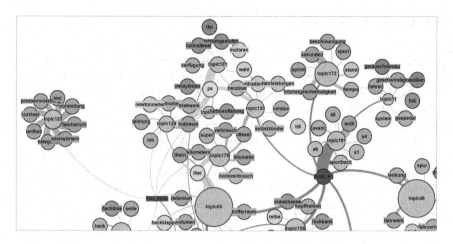

Fig. 5.21 An instance of a Topic Map (from German car reviews)

Table 5.1 Expert feedback on the Subgroup Discovery algorithm

Subgroup	Expert feedback
Activation of pro-apoptotic gene products AND regulation of apoptosis	This is an important subgroup to breast cancer. Apoptosis has been reported many times in the past (e.g. p53, BT2) for its prognostic significance (thus not novel)
Activation of pro-apoptotic gene products AND apoptosis AND signal transduction	An important subgroup too. Apoptotic effects of signal transduction pathways have been reported to metastatic breast cancer
Apoptosis AND p53 signaling pathway AND signal transduction	An important subgroup too. p53 is an important apoptosis pathway very significant for its prognostic value

interface, but the ability of the subgroups to reflect interesting knowledge in an intuitive way. Hence, focus was given to evaluate the understandability and interestingness of subgroups to experts.

In several evaluation rounds, as well as in a scientific workshop, feedback from users was evaluated. In the case of subgroup discovery on genomic data, it was shown that using highly structured features from two gene ontologies, interesting patterns could be identified. As exemplified in Table 5.1, experts could easily verify that the subgroup discovery algorithm was able to identify relevant facts about a breast cancer case (verified by comparison to the available literature).

5.6.3 Entity Prominence Service

A major outcome of Dicode was the development of a highly scalable service for Named Entity Recognition and Disambiguation (NERD). The output of the service

Fig. 5.22 An example of the
Entity Prominence Service

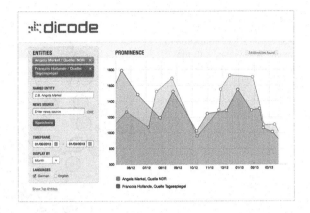

can be used by the front-end developer directly, e.g. to visualize the document
content by showing a list of named entities as a tag cloud or by giving additional
information about the entities (e.g. all persons). The service can also be used as
input for higher-level services, e.g. for topic detection. A major feature of the
service is its scalability, which allows for the annotation of huge document collections and makes it possible that the service is being used for the annotation of
significant parts of the German Internet. To visualize the service's capability of
analyzing huge document collections, the *Entity Prominence Service* was developed (Fig. 5.22). For each entity, the Entity Prominence Service depicts the
number of occurrences over time. A set of filters allows for the specification of a
news source, a time frame and the language.

5.6.4 Opinion Mining Service

This service is a combination of the *Phrase Extraction Application Service* and the
Topic Detection Service described above. It first detects phrases in a text collection. Then, it builds a topic model on those texts that contain at least one detected
phrase. In a last step, the topic model outputs the most significant sentences
assigned with each topic. This service needs a trained phrase extraction model to
operate.

The graph shown in Fig. 5.23 has been constructed from a text collection on
notebook reviews. We show a part of the whole graph that only contains topics that
are related to the query term "performance". The model was built based on a
collection of sentences that contain positive emotions. This digest expresses
(positive) opinions of consumers that are related to "performance" and to
"gaming", which is one of the main terms of this topic (topics can be defined by
words of the text collection that are related to the topic with large probabilities).

Fig. 5.23 Graphical output
of the Topic Detection
Service on a text collection of
notebook reviews with the
query term "performance"

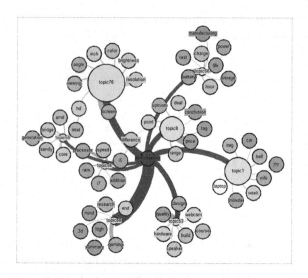

5.7 Innovative Aspects of the Dicode Data Mining Services

The key innovation of the Subgroup Discovery technology in Dicode consists of improving the applicability of the technology in the context of sense-making and decision support. While single subgroups usually offer a very high understandability because of their intuitive representation and conciseness, the understandability of a set of subgroups is limited by the fact that multiple, highly correlated subgroups can be generated. In Dicode, an improved quality criterion based on the theory of relevance was developed, which was shown to result in much more concise, less redundant sets of subgroups. Additionally, subgroup discovery was integrated into Dicode's Collaboration Support Services, thus offering objective knowledge encoded in subgroups as a direct input into a structured discussion. Finally, in the context of genomic analysis, highly structured features based on gene ontologies have been used to make subgroup discovery adaptable for the big data problem of gene analysis.

The key innovation of the Dicode Text Mining Services is to make previously existing but unconnected technologies available in one single workflow in a scalable manner. In text mining, the different steps of the text processing pipeline, such as pre-processing, entity recognition, topic detection and opinion mining cannot be perceived as independent. Instead, in order to adequately solve big data real-world problems around the question of text mining, a solution is necessary that reliably deals with large amounts of texts in an integrated way, such that the analyst receives the whole picture and not independent pieces of information.

Another major innovative aspect is the scalability of Dicode's Text Mining Services, especially of the ontology-based named entity recognition and disambiguation technique. The algorithm disambiguates named entities based on external knowledge as suggested in the literature [10]. The named entity recognition and disambiguation task can be split into multiple steps which can be easily parallelized and other steps which rely on global resources. Dicode's Named Entity Recognition and Disambiguation algorithm (NERD), implemented by Neofonie GmbH, relies on several lexical sources and ontologies which are used during different phases. The Named Entity Service was developed based on link statistics extracted from Wikipedia and a knowledge base which relies on *Freebase* (http://www.freebase.com). To exploit Hadoop's linear scaling capabilities, the developed distributed NERD (NERDist) uses Hadoop's distributed database HBase for both lexicon and ontology. The services developed in the context of Dicode run on Hadoop to meet the scalability goals. In the final phase of the project, a prototype based on *Storm* (http://storm-project.net) was developed, which not only meets the scalability requirements, but also allows for near-real-time processing and thus meets the requirement for freshness.

In the Opinion Mining Service, a set of popular text mining algorithms have been combined in an innovative way. The Topic Detection Service is based on LDA (http://en.wikipedia.org/wiki/Latent_Dirichlet_allocation), which generates topic models from text collections. The LDA training algorithm provides two tables: one that relates topics and words via probabilities, and one that relates documents and topics also via probabilities. The CRF algorithm (http://en.wikipedia.org/wiki/Conditional_random_field) is the second basic algorithm used to strengthen the text mining services in Dicode. The CRF algorithm learns to identify phrases in a sentence from examples that have been pre-annotated manually. Historically, CRF was first used to identify names of persons, organizations and geographical locations. Research in the Dicode project however revealed that the detection of more complex phrases can also be successfully trained.

Several Dicode Text Mining Services are based on the CRF algorithm: The Phrase Extraction Training Service allows training a CRF on example phrases and example texts that have been supplied by the user. The resulting CRF models have to be downloaded and stored by the user for further use with the Phrase Extraction Application Service, which extracts phrases from a text collection, using a pre-trained CRF model. The Emotion Detection Training Service can be used to train a CRF to extract phrases that convey emotions. In contrast to the Phrase Extraction Training Service, it works on a text collection and a list of so called "emotional seed terms" provided by the user. The service then extracts the actual emotional phrases from the text collection and uses them for training. The resulting CRF model can then be transferred to the Phrase Extraction Application service for use on a new text collection.

5.8 Conclusion

In Dicode, machine learning technologies—usually available to experts only—can be easily used by end users. The users can train models and share them with other users in Dicode's collaboration workspaces. The services are powerful and generic enough to be applicable to a wide spectrum of applications; moreover, they are able to deal with huge amounts of data. However, decision making on the basis of data mining relies on human expertise for the interpretation of the results obtained as well as for the customization the services for particular areas of application. The aim is to design toolboxes of customized services, based on the generic Dicode Data Mining Services, for each particular application area of Dicode.

References

1. Han, J., Pei, J., Yiwen, Y., Mao, R.: Mining frequent patterns without candidate generation: a frequent-pattern tree approach. Data Min. Knowl. Discov. **8**, 53–87 (2004)
2. Lavrac, N., Kavšek, B., Flach, P., Todorovski, L.: Subgroup discovery with CN2-SD. J. Mach. Learn. Res. **5**, 153–188 (2004)
3. van Leeuwen, M., Knobbe, A.: Diverse subgroup set discovery. Data Min. Knowl. Discov. **25**(2), 208–242. Springer, The Netherlands (2012)
4. Grosskreutz, H., Paurat, D., Rüping, S.: An enhanced relevance criterion for more concise supervised pattern discovery. In: The 18th Annual ACM SIGKDD International Conference on Knowledge Discovery and Data Mining (2012)
5. Cantone, D., Ferro, A., Pulvirenti, A., Recupero, D.R., Shasha, D.: Antipole tree indexing to support range search and k-nearest neighbor search in metric spaces. IEEE/TKDE. **17**(4), 535–550 (2005)
6. Friesen, N., Rüping, S.: Distance metric learning for recommender systems in complex domains mastering data-intensive collaboration through the synergy of human and machine reasoning (dicoSyn 2012). In: A Workshop at CSCW 2012, Seattle, WA, 12 Feb 2012
7. Thurau, C., Kersting, K., Wahabzada, M., Bauckhage, C.: Convex non-negative matrix factorization for massive datasets. Knowl. Inf. Syst. **29**(2), 457–478 (2010)
8. Trajkovsky, I.: Functional Interpretation of Gene Expression Data: Translating High-Throughput DNA Microarray Data into Useful Biological Knowledge. LAP LAMBERT Academic Publishing. (2011). ISBN: 978-3-8473-1475-2
9. Paass, G., Kindermann, J.: Entity and relation extraction in texts with semi- super-vised extensions. In: Tresp, V., Bundschus, M., Rettinger, A., Huang, Y. (eds.), Security Informatics and Terrorism: Social and Technical Problems of Detecting and Controlling Terrorists' Use of the World Wide Web; Proceedings of the NATO Advanced Research Workshop on Security Informatics and Terrorism—Patrolling the Web, vol. 15, p. 132. IOS Press, Beer-Sheva (2008)
10. Cucerzan, S.: Large-scale named entity disambiguation based on Wikipedia data. In: Proceedings of EMNLP-CoNLL, pp. 708–716 (2007)

Chapter 6
The Dicode Collaboration and Decision Making Support Services

Manolis Tzagarakis, Nikos Karacapilidis, Spyros Christodoulou,
Fan Yang-Turner and Lydia Lau

Abstract As broadly admitted, supporting collaboration and decision making in today's knowledge intensive environment is far from being easy. This is because collaboration settings are often associated with ever-increasing amounts of multiple types of data, obtained from diverse sources that often have a low signal-to-noise ratio for addressing the problem at hand. Towards addressing such concerns, we have developed a series of innovative collaboration and decision making services in the context of the Dicode project. The adopted approach facilitates and augments sense-making and decision making by incrementally formalizing the collaboration context.

Keywords Collaboration · Decision making · Incremental formalization · Data-intensiveness · Sense-making

6.1 Introduction

Current advances in computing and Internet technologies, together with the advent of the Web 2.0 era, resulted to the development of a plethora of online, publicly available environments such as blogs, discussion forums, wikis, and social

---

M. Tzagarakis (✉) · N. Karacapilidis · S. Christodoulou
University of Patras and Computer Technology Institute & Press "Diophantus", 26504 Rio Patras, Greece
e-mail: tzagara@upatras.gr

N. Karacapilidis
e-mail: nikos@mech.upatras.gr

S. Christodoulou
e-mail: shristod@cti.gr

F. Yang-Turner · L. Lau
University of Leeds, Leeds, UK
e-mail: F.Yang-Turner@leeds.ac.uk

L. Lau
e-mail: L.M.S.Lau@leeds.ac.uk

N. Karacapilidis (ed.), *Mastering Data-Intensive Collaboration and Decision Making,*   119
Studies in Big Data 5, DOI: 10.1007/978-3-319-02612-1_6,
© Springer International Publishing Switzerland 2014

networking applications. These offer people an unprecedented level of flexibility and convenience to participate in complex collaborative activities, such as long online debates of public interest about the greening of our planet through renewable energy sources or the design of a new product in a multinational company. Information found in these environments is considered as a valuable resource for individuals and organizations to solve problems they encounter or get advice towards making a decision.

At the same time, today's knowledge work is collaborative in nature [1–3]. In many fields, such as in bioinformatics and marketing, multidisciplinary teams are formed and collaborate in order to confront complex problems. When such teams are engaged in collaborative activities, they usually have to go through some type of sorting, filtering, ranking and aggregation of the existing resources in order to facilitate sense- and decision making. Yet, these activities are far from being easy. This is because collaboration settings are often associated with ever-increasing amounts of multiple types of data, obtained from diverse sources that often have a low signal-to-noise ratio for addressing the problem at hand [4]. In turn, these data may vary in terms of subjectivity, ranging from individual opinions and estimations to broadly accepted practices and indisputable measurements and scientific results. Their types can be of diverse level as far as human understanding and machine interpretation are concerned. They can be put forward by people having diverse or even conflicting interests. At the same time, the associated data are in most cases interconnected, in a vague or explicit way. Data and their interconnections often reveal social networks and social interactions of different patterns.

In the context of collaboration and decision making support, the above bring up the need for innovative software tools that can appropriately capture, represent and process the associated data and knowledge, while at the same time remedy the underlying cognitive overload issues. Such tools should shift in focus from the collection and representation of information to its meaningful assessment and utilization. They should facilitate argumentation (i.e. discussion in which reasoning and disagreements exist, not only discourse for persuasion, logical proof and evidence-based belief [5]), the ultimate aim being to augment collaborative sense-making and/or decision-making. This can be seen as a special type of social computing where various computations concerning the associated context and group's behavior need to be supported.

While contemporary collaboration and decision making support tools are helpful in particular settings, current solutions prove to be inadequate in addressing the requirements of multidisciplinary teams working in knowledge intensive environments (see Chap. 2). In this chapter, we present the approach taken in the context of the Dicode project to support collaboration and decision making. The proposed solution is capable of tackling the diversity and complexity of the above issues, the ultimate goals being to make it easier for users to follow the evolution of an ongoing collaboration, comprehend it in its entirety, and meaningfully aggregate data in order to resolve the issue under consideration.

The chapter is organized as follows: first, we present requirements and challenges related to supporting collaboration and decision making in knowledge intensive environments. We then present the view-based approach taken in Dicode to support collaboration and decision making and discuss some key aspects. Next, we present CommBAT, a tool that allows the monitoring and investigation of the collective behavior of teams with respect to sense-making tasks. A discussion on how Dicode's collaboration and decision making support services fulfill the abovementioned requirements concludes the chapter.

6.2 Requirements and Challenges

To meet the challenges associated with supporting collaboration and decision making in the context of Dicode, we performed a series of interviews to identify the major issues that stakeholders face during their collaboration practices. These were:

- **Information overload**. This is primarily due to the extensive and uncontrolled exchange of diverse types of data and knowledge resources. For instance, such a situation may appear during the exchange of numerous ideas about the solution of a public issue, which is accompanied by the exchange of big volumes of positions and arguments in favor or against each solution.
- **Difficulty in monitoring social behavior**. The representation and visualization of social structures, relationships and interactions taking place in a collaborative environment with multiple stakeholders are also of major importance. This is associated to the perception and modeling of actors, groups and organizations and their behaviors in the diversity of collaborative contexts. A problem to be addressed is to provide the means to appropriately represent and manage user and group profiles, as well as social relationships given that they are not static but changing over time.
- **Diversity of collaboration modes**. Interviews indicated that the evolution of a collaboration session proceeds incrementally; ideas, comments, or any other type of collaboration objects are exchanged and elaborated, and new knowledge emerges slowly. When members of a community participate in a collaborative session, enforced formality may require them to specify their knowledge before it is fully formed. Such emergence cannot be attained when the collaborative environment enforces a formal model from the beginning. On the other hand, formalization is required in order to ensure the environment's capability to support decision making or estimate the present state of the collaboration.
- **Expression of tacit knowledge**. A group of people is actually an environment where tacit knowledge (i.e. knowledge that the members do not know they possess or knowledge that members cannot express with the means provided) predominantly exists and dynamically evolves.
- **Difficulty in exploiting and integrating legacy resources**. Many resources required during a collaborative session have either been used in previous sessions or reside outside the members' working environment such as e-mails and

results from the execution of various data processing algorithms. Moreover, outcomes of past collaboration activities should be able to be reused as input in subsequent collaborative sessions. Such functionality must be provided in ways that do not disrupt or impede an ongoing collaboration.

- **Data processing and decision making support**. In the settings under consideration, timely processing of data related to both the social context and social behavior is required. Such processing will significantly aid the members of a community to conclude the issue at hand (i.e. extract meaningful knowledge and reach a decision). This means that their environment needs to interpret the knowledge item types and their interrelationships in order to proactively suggest trends or even aggregate data and calculate the outcome of a collaborative session.

The above issues delineated some categories of crucial requirements to be met during the development of Dicode's collaboration and decision making support services.

6.3 The Dicode Approach

Support for collaboration and decision making in Dicode brings together two paradigms: the *Web 2.0 paradigm*, which builds on flexible rules favoring ease-of-use and human interpretable semantics, and the *traditional decision support paradigm*, which requires rigid rules that reduce ease-of-use but render machine interpretable semantics. To achieve this, our approach builds on a conceptual framework, where formality and the level of knowledge structuring during collaboration is not considered as a predefined and rigid property, but rather as an adaptable aspect that can be modified to meet the needs of the tasks at hand. By the term formality, we refer to the rules enforced by the system, with which all user actions must comply. Allowing formality to vary within the collaboration space, *incremental formalization*, i.e. a stepwise and controlled evolution from a mere collection of individual ideas and resources to the production of highly contextualized and interrelated knowledge artifacts and finally decisions, can be achieved [6].

Dicode offers alternative visualizations of the collaboration workspace (called Dicode views), which comply with the above mentioned incremental formalization concept. Each Dicode view provides the necessary mechanisms to support a particular level of formality. The more informal a view is, the greater easiness-of-use is implied. At the same time, the actions that users may perform are intuitive and not time consuming; however, the overall context is human (and not system) interpretable. On the other hand, the more formal a view is, the smaller easiness-of-use is rendered; the actions permitted are less and less intuitive and more time consuming. The overall context in this case is both human and system interpretable [7]. The views supported in our approach are:

- **Discussion-Forum view**, where a collaboration workspace is displayed as a traditional web-based forum, where posts are displayed in ascending chronological order. Users are able to post new messages to the collaboration space, which will appear at the end of the list of messages. The aim of this view is to allow the collection and sharing of opinions without limiting the expressiveness of participants.
- **Mind-map view**, where a collaboration workspace is displayed as a mind map that enables an informal representation and interrelation of collaboration items, while bearing a set of useful semantics.
- **Neighborhood view**, which allows users to select one particular item on the Mind-map and display only those items that are directly related to it, in order to focus on the context of the selected item.
- **Formal view**, which adheres to a specific argumentation model (i.e., IBIS [5]) and invokes a set of dedicated scoring and reasoning mechanisms to aid users conceive the outcome of a collaborative session and receive support towards reaching a decision.
- **Multi-criteria decision making view**, where a set of multi-criteria decision making algorithms can be executed to rank the alternative solutions.

During collaboration sessions, each user can individually choose the view with which he/she may want to conduct the collaboration. In the following, we present the last four views in greater detail.

6.3.1 The Mind-Map View

In this view, the collaboration workspace is displayed as a mind map (Fig. 6.1), where users can upload and interrelate diverse types of items. This view deploys a spatial metaphor permitting the easy movement, arrangement and structuring of items on the collaboration workspace. The aim of this view is to support *information triage* [8], i.e. the process of sorting and organizing through numerous relevant materials and organizing them to meet the task at hand.

While working in the Mind-map view of the collaboration workspace, stakeholders may organize their collaboration through dedicated item types such as ideas, notes, comments and services. *Ideas* stand for items that deserve further exploitation; they may correspond to an alternative solution to the issue under consideration and they usually trigger the evolution of the collaboration. *Notes* are generally considered as items expressing one's knowledge about the overall issue, an already asserted idea or note. *Comments* are items that usually express less strong statements and are uploaded to express some explanatory text or point to some potentially useful information. Finally, *service* items enable users to upload, configure, trigger and monitor the execution of external services from within the collaboration workspace, and allow the automatic upload of their results into the workspace (as soon as the execution of the service is completed). The service

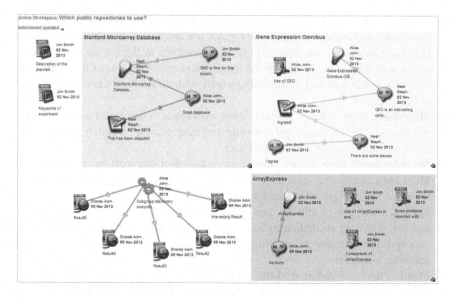

Fig. 6.1 A collaboration workspace in the mind-map view

items as well as the results they produce are part of the discourse and can be handled like any other of the available items. Multimedia resources can also be uploaded into the Mind-map view (the content of which can be displayed upon request or can be directly embedded in the workspace). In any case, the set of available item types in the Mind-map view is not fixed; users may expand the existing set by creating new types to be used during their collaboration. This allows them to tailor the discourse to the needs of the problem at hand. Users may rate individual items on a 1–5 scale indicating the importance of each item.

All item types can be explicitly related to express agreement, disagreement, support, request for refinement, contradiction etc. Visual cues are used to indicate the semantics of such relationships: for instance, a green-colored relationship indicates agreement, while a red-colored one indicates disagreement. Moreover, the thickness of a relationship may express how strongly an item agrees with or objects to another one. Finally, the Mind-map view provides abstraction mechanisms that enable items to be aggregated and be treated as a single entity within the workspace (see the colored rectangles in Fig. 6.1).

6.3.1.1 Using Service Type Items

Whenever participants need to execute external services (e.g. data mining services), they can use items of the "service" type. Such items enable external services to upload their outcomes into the collaboration workspaces, thus making such entities part of the discourse. A REST-based approach for integrating external services with the collaboration workspace has been adopted.

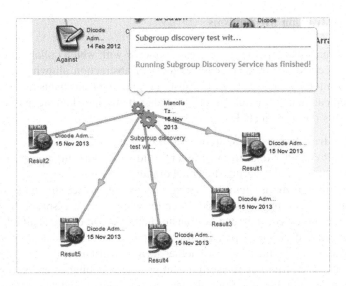

Fig. 6.2 A service item on the collaboration workspace has finished the execution of the subgroup discovery service. The outcomes of the execution are automatically uploaded and related in the workspace (items "Result1"–"Result5")

After uploading such an item into the workspace, users may configure it by specifying the URI for the REST-based external service along with all the necessary parameter required by the service to be executed. Once the service item has been configured, users may start its execution by double clicking on it. A visual cue on the item allows users to monitor the status of its execution: service items appear grey if they have not been configured, green when the execution is still ongoing, orange if the execution has finished successfully, and red if an execution error has occurred. When the execution of the associated external service terminates, its outcomes are automatically added into the collaboration workspace. Figure 6.2 shows a service item on the collaboration workspace that has successfully terminated its execution. The service executed in Fig. 6.2 implements the Subgroup Discovery algorithm [9], a popular technique in the bioinformatics field which allows finding subgroups of a population with high generality and distributional unusualness with respect to a target attribute (see Chap. 5). As shown in Fig. 6.2, five items are connected to the service item (labeled "Result1"–"Result5"), which correspond to the service outcomes, i.e. the discovered subgroups. The outcomes of external services are uploaded into the collaboration workspace as XML documents.

As is the case for any collaboration item, users can collaboratively discuss the obtained results by uploading items that argue in favor or against them. Moreover, taking into consideration the discussion about the results, users may decide to re-execute the external service with different parameters.

6.3.1.2 Abstraction Mechanisms

The Mind-map also provides the necessary means with which collaboration items can be conceived at a higher or lower level of abstraction allowing their transformation into new artifacts. These constitute important mechanisms to facilitate the piecemeal transformation of the available items into knowledge structures. Mechanisms provided include:

- **Explicit transformation of items**. Individual items can be transformed from one type to another (e.g. from comment to idea) without any constraint by any user and at any point during the collaboration.
- **Aggregation of items**. Items can be aggregated into larger structures and these structures can be treated as a single entity within the workspace. Aggregated items may be typed; they can be transformed into one of the available knowledge types and can take part in any structuring activity, such as relating an aggregated entity with a note or another idea. For instance, a set of aggregated items can be cast into an idea, comment or note. Undoing of an aggregation is also possible. In these situations, the aggregation is dissolved and the constituent parts appear as separate items on the collaboration space.
- **Breaking-up of items**. Individual items may be broken up into smaller pieces in order to allow these smaller pieces to take part in relationships (Fig. 6.3). In particular, a part of an item's content may be selected and be treated as a separate item in order to be more precise during an argumentative collaboration. During such a break up, the relationship to the original item is maintained as an attribute of the new instance, so that its origin can be traced back.
- **Patterns of knowledge structures**. Instances of interconnected knowledge items (of any type) can be designated as knowledge type templates. These templates can then be used during the collaboration. This allows the definition and use of user-defined abstractions during the collaboration.

6.3.1.3 Filtering Collaboration Workspaces

The Mind-map view provides an additional functionality that allows users to filter out items that appear on the collaboration workspaces. Such filtering permits users to keep only a subset of the available items on the workspace, while hiding all the others. The purpose of this functionality is to allow users to focus on particular items on the workspace when the workspace becomes large, with many interrelated items. Its overall aim is to make the discourse more understandable to participants when the number of items grows large.

A criteria-based approach has been adopted to filter out items on a workspace: users may specify what criteria the items that they want to keep on the workspace should match. Figure 6.4 shows the interface that enables users to specify the filtering criteria.

Users can filter collaboration workspaces based on the following criteria:

Fig. 6.3 Example of breaking-up an item. The selected part of the content can be used as a separate item with a distinct type within the workspace

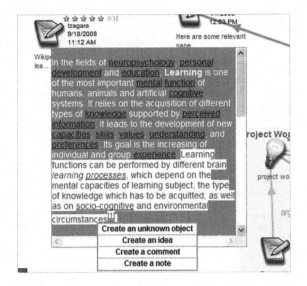

- The creator of items, which allows one to specify items that were uploaded by a specific user.
- The last user modifying the item, which allows to specify items that were last modified by a particular user (in the Mind-map view, all participants can modify the available items).
- The item's Mime-type, which allows users to specify which Mime-type the items should match. This allows users to specify Mime-types such as pdf, doc or ps files.
- The item's knowledge type, which allows users to keep on the workspace items which are of a particular knowledge type.
- The item's creation date, which allows users to keep items that have been modified within a specific period.
- The item's argumentation level, which allows one to keep items that are connected to an item of knowledge type "idea" via a path whose length is less or equal to a specified value. In the Mind-map view, items can be explicitly associated via arrows in order to express a particular semantic relationship. Considering such arrows as vertices and items as nodes, the mind map can be conceptualized as a graph. The terms "path" and "length" is perceived in the graph theoretic sense.

The above criteria are used conjunctively; items matching all specified criteria are kept on the collaboration workspace, while items not matching even one of them are temporarily removed. Such removal is not permanent; during filtering, these items and their relationships are just not shown on the collaboration workspace. When the user exits the filtered view, the filtered out items will show up again.

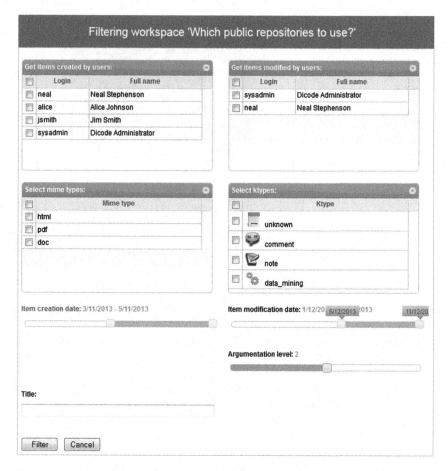

Fig. 6.4 Specifying the filtering criteria of a collaboration workspace

6.3.1.4 Sharing Collaboration Workspaces

Collaboration workspaces can be shared via popular social networking sites. In particular, this functionality allows users to post the URL of collaboration workspaces into social networking sites, through which access to the shared workspace is possible. Such a sharing is supported for many popular social networking sites (including Facebook, Twitter, LinkedIn, etc.) by exploiting the relevant APIs that these sites provide. A main menu option gives users access to this functionality (Fig. 6.5).

Fig. 6.5 Sharing collaboration workspaces with popular social networking sites

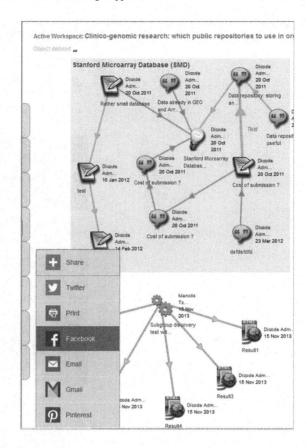

6.3.2 The Neighborhood View

The Neighborhood view of a collaboration item displays the specified item along with its neighborhood. The "neighborhood" of a specific collaboration item is defined as the set of items with which that item is directly connected via a relation in the Mind-map view. The aim of this view is to allow users focus on immediately connected items and not be distracted by others.

Figure 6.6 shows the Neighborhood view of a selected collaboration item. The selected item is depicted on the top of the view, while the item's neighborhood is depicted beneath it with a colored background depending on the type of relation they bear with the item (green for "in favor", red for "against", grey for "neutral"). The view also supports a number of operations. In particular, users may "like" or "dislike" all displayed items or update and even create a new one. The total number of "likes" and "dislikes" received is depicted for each collaboration item (textually and graphically through a colored bar), while all users who "liked" or "disliked" are also viewable. The user may also move to another item's neighborhood (by

Fig. 6.6 Neighborhood view of a collaboration item

moving it up on the top level of the hierarchy) or choose to load all workspace items and their respective neighborhoods.

6.3.3 The Formal View

The aim of the Formal view is to support users in reaching a decision during the collaborative session. It provides an argumentative discourse with a fixed set of discourse element and relationship types, with predetermined, system-interpretable semantics. This view adopts an IBIS-like formalism; the item types supported are *issues*, *alternatives* and *positions*. It provides a structured language for argumentative discourse together with a mechanism for the evaluation of alternatives. Additional reasoning can be performed through the expression of *preferences*, which provide participants with a qualitative way to weigh reasons for and against

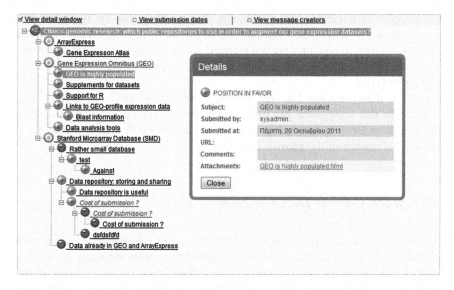

Fig. 6.7 The formal view of the collaboration workspace shown in Fig. 6.1

Table 6.1 Set of rules that transform a collaboration workspace from the Mind-map view into the formal view

Type in the mind-map view	Type in the formal view
Title of the collaboration workspace	Issue
Idea	Alternative
Relationship between comment/note/service and idea colored red	Position against an alternative
Relationship between comment/note/service and idea colored green	Position in-favor of an alternative
Relationship between comments/notes/services colored red	Position against another position
Relationship between comments/notes/services colored green	Position in-favor of another position
Thickness of relationships	Weight of the position

the selection of an alternative. Further to the argumentation-based structuring of a collaborative session, the Formal view integrates a reasoning and scoring mechanism (for details, see [10]), which determines the status of each collaboration item, the ultimate aim being to keep users aware of the most prominent alternative.

Figure 6.7 shows the Formal view of the collaboration workspace depicted in Fig. 6.1. By clicking on a position or alternative in the Formal view, more details related to the selected item can be displayed (see the "Details" window in Fig. 6.7).

A rule-based approach facilitates the transformation of a collaboration workspace operated in the Mind-map view into the Formal view. These rules specify how the item types in the Mind-map will appear in the Formal view when the collaboration workspace is transformed. Table 6.1 summarizes the set of rules which enable such transformation. Visual cues in the Mind-map are also taken into consideration. In particular, items in the Mind-map view connected with red arrows to an item of type idea are transformed into positions arguing against; items connected with a green arrow to an item of type idea are transformed into positions arguing in favor.

After a transformation into the desired view occurs, the collaboration may continue in this view, with the users being able to exploit the semantic types available in order to keep conducting the discourse in the desired formality level and take advantage of the provided services.

6.3.4 Multi-Criteria Decision Making View

The Multi-Criteria Decision Making (MCDM) view of a collaboration workspace is a read-only view; its main purpose is to further support the decision making process by considering the attributes of the collaboration items appearing in the Mind-map view and exploiting diverse MCDM algorithms to indicate prevailing solutions. Based on the attributes of each alternative, each MCDM algorithm calculates a corresponding alternative score; the alternative with the highest score is considered to be the best solution to the problem at hand. In Dicode, four attributes/criteria are used for the evaluation of each alternative:

- **Likes/Dislikes**. The algebraic sum of an item's number of "Likes" minus its number of "Dislikes".
- **Creator rating**. Calculated as the algebraic sum of all "Likes" minus all "Dislikes" corresponding to the items the creator has contributed on a workspace.
- **Relationships in favor/against**. The algebraic sum of an item's number of "in favor" relationships (depicted with green arrows in the Mind-map view) minus the item's number of "against" relations (depicted with red arrows in the Mind-map view).
- **Item rating**. The total rating corresponding to the users' preferences (expressed through a 1–5 rating scale).

The selection of the MCDM algorithms to be implemented in the context of this view was based on a questionnaire filled in by senior decision makers, acting in diverse data-intensive settings. According to the results of this questionnaire, the best suited decision making methodology highly depends on the specific problem under consideration. Depending on the specific problem, decision makers would require support from methodologies that: (i) allow compensation among the attributes/criteria used for the evaluation of the alternatives (i.e. a good performance of an alternative concerning one attribute can compensate for a bad performance concerning another attribute), (ii) allow two or more alternatives to be incomparable, and (iii) do not allow compensation among criteria.

Three MCDM algorithms, fulfilling the aforementioned prerequisites, have been implemented in the context of this view (these algorithms are briefly presented at

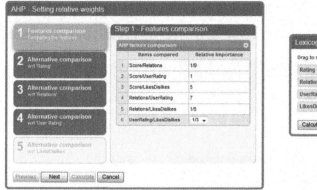
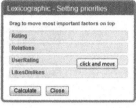

Fig. 6.8 Forms for setting the relative weights in AHP (*left*) and sorting the attributes/factors with respect to their importance in LDM (*right*)

the end of this section): the *Weighted Sum Model* (WSM) [11], the *Analytical Hierarchy Processing* (AHP) [12] and the *Lexicographic Decision Making rule* (LDM) [13]. For each algorithm, the user has to set the necessary parameters. Upon the execution of the algorithm, the calculated ranked list of the alternatives is returned. Figure 6.8 shows the forms that allow users to configure the AHP and LDM algorithms. After configuring and executing the desired algorithm, the user may browse through the detailed results of the algorithm (to realize the reason why an alternative performs better than another one), view the plot with the scores of the alternatives or reset the algorithm's parameters to perform a 'what-if' (sensitivity) analysis [14]. The mechanisms developed in this view build on the reasoning capabilities of the machine to enhance decision making. Figure 6.9 shows the MCDM view of the collaboration workspace depicted in Fig. 6.1; the ranking of the alternatives for each implemented algorithm is included.

6.3.4.1 The Weighted Sum Model

The Weighted Sum Model (WSM) is the most popular MCDM approach. For a number of *M* alternatives, the best alternative is the one with the top score calculated as:

$$A^*_{wsm} = max_i \sum_{j=1}^{N} q_{ij}w_j \quad for\, i = 1, 2, 3, \ldots, M$$

where A^*_{wsm} is the score calculated for the best alternative, N is the number of criteria (for the Dicode case, $N = 4$), q_{ij} is the subscore of the i-th alternative with respect to the j-th criterion, and w_j is the (user-defined) weight reflecting the relative importance of the j-th criterion.

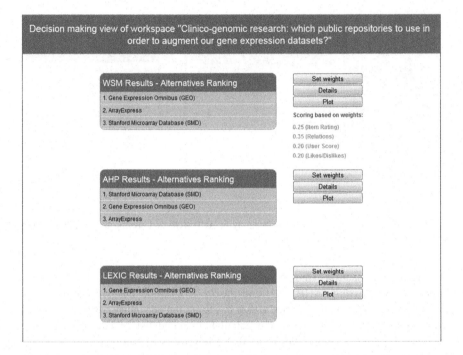

Fig. 6.9 The multi-criteria decision making view of a collaboration workspace

6.3.4.2 Analytic Hierarchy Process

Analytical Hierarchy Processing (AHP) is based on decomposing a problem into a system of hierarchies. Calculation of the alternatives ranking relies on pair-wise comparisons among the criteria and among the offered alternatives (with respect to each criterion). Two or more alternatives are allowed to be incomparable. In the context of Dicode, a wizard is used to set the relative importance values for all pairs of criteria and alternatives.

6.3.4.3 The Lexicographic Decision Making Rule

The Lexicographic Decision Making (LDM) rule is a decision rule based on ranking the attributes of the decision making process on terms of their importance. No compensation is allowed between the attributes. In the context of Dicode, the user has to rank the four attributes (criteria) based on their importance. The calculation of the rank of alternatives is based on the partial score (i.e. the performance) of each alternative with respect to the most important attribute.

6.4 Monitoring Collaborative Behavior

Dicode's collaboration and decision making support services provide not only the necessary infrastructure to facilitate brainstorming and decision making in data intensive contexts, but also offer mechanisms to monitor and investigate the collective behavior of teams with respect to sense-making tasks. In general, a key challenge when using online tools for collective sense-making is to gain an insight of the collective user behavior and identify situations when a moderator should interfere to steer the collaboration process. The overall aim of these mechanisms in Dicode is to understand the sense-making processes by addressing questions such as:

- How does a group make collectively sense of a problem in the supported collaboration workspaces in Dicode?
- What kind of sense-making behaviors are involved and can be witnessed in collaboration workspaces?
- Is it possible to discover any patterns within the discourse which may lead to automatically facilitate the collaborative process from brain storming to decision making?

Towards addressing such challenges, we have developed *CommBAT* (Community Behavior Analytics Tool), a desktop application, which presents statistics of Dicode collaboration workspaces' log data in a visual form. It enables a user to dynamically explore facets captured by the log data: participants, objects (ideas, comments, notes or services), activities (view, create, update or move objects) and semantic types (relationships between objects and activities).

CommBAT is designed with the following users in mind: (i) developers of Dicode collaboration and decision making support service who may be interested in how users benefit from the novel features of the system and how to support them better; (ii) researchers in collaboration and decision making studies who may be interested in identifying patterns of user and community behavior from the interaction logs of Dicode collaboration workspaces; and (iii) community leaders or moderators who may be interested in how to use the tool to improve the process of collaboration and decision making.

CommBAT provides three features: (i) *Dynamic Filtering*—the user can select facets to be extracted from the log data, (ii) *Visual Presentation*—results can be displayed in different types of chart: pie, column, bar, bubble etc.; and (iii) *Multiperspective views*—users can conduct investigations into knowledge items (ideas, comment, notes and services) using object type view, user activity view and/or object timeline view.

Figure 6.10 demonstrates a possible output screen for analysis. The three mini charts across the top from left to right are: (i) *User information*—shown in a column chart, with the participants with their number of activities sorted in a descendent order; (ii) *Sense-making actions*—a pie chart which summarizes what users had performed in the workspace: contributing, reviewing or organizing

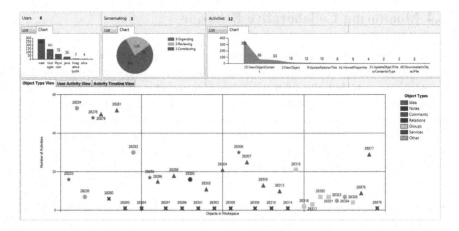

Fig. 6.10 CommBAT with Views of Users, Sensemaking Actions, Workspace Activities and Object Type

knowledge items etc.; and (iii) *Workspace activities*—showing details of sense-making actions such as how users contributed to the workspace, creating new knowledge objects or new relations and so on. The bottom half of Fig. 6.10 shows *Object Type View*, in which the frequency of each object when active is displayed along the y-axis and the creation date along the x-axis. Each object type is coded in a different shape and color. For example, one observation from this chart is: users started to group objects (in pink squares) towards the later part of the workspace's lifespan, while relationships between items (in blue crosses) are constantly created during the whole lifespan.

CommBAT provides a Multiple Document Interface (MDI), which supports multiple logs of workspaces presented at the same time. This allows users to compare different analytic results among different workspaces. For example, the screen shot in Fig. 6.11 compares two workspaces with ID 29967 and ID 27231. CommBAT also provides *User Activity View* (UAV) and *Activity Timeline View* (ATV). A detailed description of all related features of CommBAT appears at https://sites.google.com/site/commbatwiki/.

CommBAT enables visual, dynamic log analysis to examine the community features of the Dicode collaborative workspaces such as participants and activities. For a user who looks for patterns in sense-making activities, this tool opens up a pool of objective data captured during user interactions for further analysis. For example:

- How participants contribute to the workspace in terms of activities?
- How participants use the objects to contribute to the workspace?
- How an object evolves in terms of its interaction with participants and over a timeline?
- How participants interact with one type of object over a timeline?
- What interactions have taken place surrounding a type of object or an object?

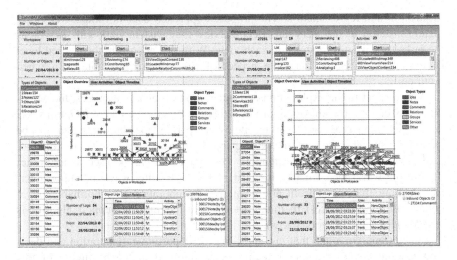

Fig. 6.11 Using CommBAT to compare two workspaces

With CommBAT, user and community behavior can be discovered so that the quality and evolution of collaboration and decision making can be monitored. This is a step forward towards further improving the Dicode collaboration and decision making support services.

6.5 Discussion and Concluding Remarks

The Dicode collaboration and decision making support services provide functionalities that help alleviate the impact of data-intensiveness during collaboration sessions. With respect to such issues, the Mind-map view offers ways to structure the discourse, generate new abstractions by aggregating items as well as filtering a workspace in order to allow participants to focus only on specific comments and alternatives. Such mechanisms enable users to control the complexity of collaboration workspaces while at the same time making the discourse understandable.

As far as monitoring social behavior is concerned, the CommBAT framework allows the examination of the collective behavior of teams through which important conclusions related to their sense-making actions can be drawn. Furthermore, it enables drawing useful conclusions related to the profiles of users and groups taking into consideration their dynamic nature.

The view-based approach to supporting collaboration provides a diverse range of collaboration modes which enables the incremental formalization of collaboration workspaces; this permits the evolution of the discourse items from loosely coupled artifacts into new knowledge items and finally decisions. The use of alternative views and the ability to shift between them enhances tacit knowledge acquisition and representation. The proposed approach does not impose any premature structure as the users can select the view they wish to work with, as well as the tasks they want to perform in each view. With this approach, the desired level of formalization is an adaptable aspect of the system, available only when required.

With respect to the exploitation and integration of legacy resources, the Mindmap view enables the uploading of items that represent external computations, via the service item type. Such service items are part of the discourse and can be treated as any other item on the workspace. Moreover, their results are automatically uploaded into the corresponding workspace and can be used as any other item of the workspace. Finally, the designed services offer ways to support decision making that range from argumentative collaboration (the Formal view of collaboration workspaces) to multi-criteria decision making support algorithms (the MCDM view of workspaces).

Concluding, we argue that our approach to supporting collaboration and decision making covers the set of requirements outlined in the beginning of this chapter. The Dicode approach offers the proper foundation for addressing issues related to the varying levels of formality needed in collaborative knowledge building within multidisciplinary teams working in knowledge intensive settings.

References

1. Brazas, M.D., Tadashi, J.T. Yamada, Ouellette, F.B.: Evolution in bioinformatic resources: 2009 update on the Bioinformatics links directory. Nucleic Acids Res. **37**, W3–W5, 20 (2009)
2. Hara, N., Solomon, P., Kim, S.-L., Sonnenwald, D.H.: An emerging view of scientific collaboration: Scientists' perspectives on collaboration and factors that impact collaboration. J. Am. Soc. Inform. Sci. Technol. **54**, 952–965 (2003)
3. Sally Lee, E.: Facilitating collaborative biomedical research. In: ACM GROUP '07 Doctoral Consortium Papers, pp. 5:1–5:2. New York, USA (2007)
4. Eppler, M., Mengis, J.: The concept of information overload: a review of literature from organization science, accounting, marketing, MIS, and related disciplines. Inf. Soc. **20**(5), 325–344 (2004)
5. Kunz, W., Rittel, H.: Issues as elements of information systems. Technical report 0131, Institut für Grundlagen der Planning, Universität Stuttgart (1970)
6. Shipman, F.M., McCall, R.: Supporting knowledge-base evolution with incremental formalization. In: Proceedings of CHI 94 Conference, pp. 285–291 (1994)
7. Karacapilidis, N., Tzagarakis, M.: Towards a seamless integration of human and machine reasoning in data-intensive collaborative decision making settings: the dicode approach. In: Proceedings of the 16th IFIP WG8.3 International Conference on Decision Support Systems (DSS 2012), pp. 223–228. IOS Press, Amsterdam (2012)
8. Marshall, C.C., Shipman III, F.M.: Spatial hypertext and the practice of information triage. In: Proceedings of the 8th ACM conference on Hypertext, April 06–11, pp. 124–133. Southampton, UK (1997)
9. Grosskreutz, H., Paurat, D., Rueping, S.: An enhanced relevance criterion for more concise supervised pattern discovery. In: 18th Annual ACM SIGKDD International Conference on Knowledge Discovery and Data Mining (2012)

10. Karacapilidis, N., Papadias, D.: Computer supported argumentation and collaborative decision making: the HERMES System. Inf. Syst. **26**(4), 259–277 (2001)
11. Fishburn, P.C.: Additive Utilities with Incomplete Product Set: Applications to Priorities and Assignments. Operations Research Society of America (ORSA), Baltimore (1967)
12. Saaty, T.: The Analytic Hierarchy Process: Planning, Priority Setting, Resource Allocation. McGraw-Hill, New York (1980). ISBN 0-07-054371-2
13. Fishburn, P.C.: Lexicographic order, utilities and decision rules: a survey. Manage. Sci. **20**, 1442–1471 (1974)
14. Triantaphyllou, E., Sanchez, A.: A sensitivity analysis approach for some deterministic MCDM methods. Decis. Sci. **28**(1), 151–194 (1997)

Chapter 7
Integrating Dicode Services: The Dicode Workbench

Guillermo de la Calle, Eduardo Alonso-Martínez, Martha Rojas-Vera and Miguel García-Remesal

Abstract This chapter presents the innovative approach developed in the Dicode project regarding the integration of services and applications in the context of "big data". The objective was to define a flexible, scalable and customizable information and computation infrastructure to exploit the competences of stakeholders and information workers by incorporating the underlying collective intelligence. Our approach pays much attention to usability and ease-of-use issues, aiming to enable users without any particular programming expertise to use the system. We present two major outcomes of the Dicode project regarding integration issues: the Dicode Workbench and the Dicode Integration Framework.

Keywords Service integration · Service architectures · SOA · Mashup application · Data-intensiveness · Collaborative work · Collective intelligence

7.1 Introduction

The main objective of the Dicode project regarding integration was to develop a flexible software architecture, customizable for different scenarios (use cases), to allow the integration of already existing services with new services specifically

G. de la Calle (✉) · E. Alonso-Martínez · M. Rojas-Vera · M. García-Remesal
School of Computer Science, Universidad Politécnica de Madrid, Madrid, Spain
e-mail: gcalle@infomed.dia.fi.upm.es

E. Alonso-Martínez
e-mail: ealonso@infomed.dia.fi.upm.es

M. Rojas-Vera
e-mail: mrojas@infomed.dia.fi.upm.es

M. García-Remesal
e-mail: mgarcia@infomed.dia.fi.upm.es

N. Karacapilidis (ed.), *Mastering Data-Intensive Collaboration and Decision Making*,
Studies in Big Data 5, DOI: 10.1007/978-3-319-02612-1_7,
© Springer International Publishing Switzerland 2014

developed within the project. The overall development process exploited existing open source frameworks, toolkits and libraries, while much attention was given to reusability, scalability and expandability issues. Another objective was to develop an integration framework to facilitate the coordination and interoperability of Dicode services. This integration had to be carried out from both a conceptual and a technical point of view. This means not only to provide users with an integrated environment to use different services but also to offer a complete integration framework where services can interoperate.

From the beginning of the project, all Dicode partners agreed on developing light-weight services which could be integrated via a REST-based interface or directly as a widget into the Dicode Workbench. This pattern does not pose any restrictions on the back-end technology used for service development. In Dicode, components of the system are wrapped into services and integration is performed on a service level. Due to the abovementioned agreement, the need for common coding and team organization practices was simplified.

In the next sections, the Dicode Workbench and the Dicode Integration Framework are described in detail. Guidelines and instructions for developing and integrating services into the Dicode Workbench are presented at the end of the chapter.

7.2 The Dicode Workbench

The Dicode Workbench has been conceived as a web application to provide users with a common graphical interface to access and use heterogeneous services. The Dicode Workbench is the integration platform for all Dicode data analysis and collaboration services. After analyzing the different possibilities available, the Dicode consortium decided to adopt a web-based solution for the following reasons:

- Web applications are worldwide accessible just using a computer with a web browser and Internet connection;
- Users do not need to install any extra application on their computer. Therefore, no problem with virus or malware applications will arise;
- Web applications are platform and operating system independent;
- Security problems are avoided since most web applications are firewall friendly, using common ports that usually are not filtered by security policies of organizations. That way, applications availability is ensured;
- Different users can easily and directly share workspaces and applications.

As most web applications, the Dicode Workbench has two different views: the public and the private one. The public view is accessible by any user with Internet connection. It contains general information of the Workbench and allows users to register in the system. The private view is only available for registered users.

Since web applications execute within web browsers, it is crucial that they can properly run on the most popular ones. In the Dicode project, we paid much attention to this issue to ensure that users have all functionalities available independently from the operating system or web browser used. Specifically, we considered five of the most popular web browsers, namely: Internet Explorer, Mozilla Firefox, Google Chrome, Safari and Opera. The Dicode Workbench has been implemented using Java technologies, i.e. JavaServer Pages (JSPs) and Servlets [1]. It is publicly available at http://hodgkin.dia.fi.upm.es:8080/dicode. The basic features of the Dicode Workbench are presented below.

7.2.1 Site Map

Figure 7.1 shows the site map of the Dicode Workbench. As stated above, all pages have been implemented using JavaServer Pages (JSP) technology [1]. Arrows coming from a web page denote the different options that users have available on a page. White rectangles specify those options. In this graphic, a special structure has been defined called "Confirmation/Error Schema" (C/E Schema). This schema represents the possible results coming from some pages. When users complete an action, this action may finish with success or with an error. C/E Schema models these situations, showing a confirmation page when an action finishes successfully or an error page otherwise.

The initial page (index.jsp) shows a welcome message and allows users to login into the Dicode Workbench. Departing from this page, the application web pages of the Dicode Workbench can be grouped into three main categories according to their functionalities:

- **Unregistered user pages**. Pages that users can access without registration compose this category. There are three pages:
  - *Index.jsp*—the initial page as described above.
  - *Registration.jsp*—this page allows users to create a new user in the Dicode Workbench.
  - *Remember.jsp*—this page allows users to recover their password to access the system whenever they forgot it.
- **Registered user pages**. This category refers to the options that users have after a successful login process. These options are displayed inside a menu box on the upper left side of the page.

  - *Welcome.jsp*—this is the initial page when users login into the system. General information about the Workbench and the latest news are displayed.
  - *Profile.jsp*—this page allows users to display and update their personal information in the system.
  - *Info.jsp*—this page shows general information about the Dicode project: main objectives, partners, contact information, etc.
  - *Services.jsp*—this page presents information about the services that service providers have published in the Dicode Workbench.

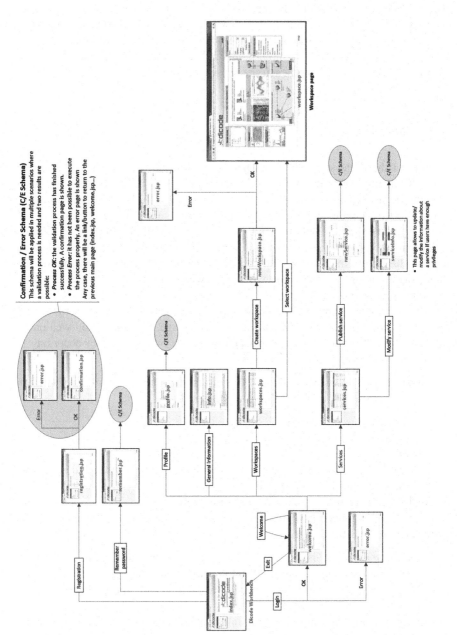

Fig. 7.1 Dicode workbench site map

Welcome to the Dicode Workbench

The Dicode workbench is an advanced web application enabling the integration of multiple tools under the same interface. It is based on a flexible software architecture to facilitate the adaptation of the user interface to different user communities and types of activities.

The Dicode workbench allows

- using Dicode's tools and services in a web-based solution customizable the specific needs of the different scenarios (biological, medical, opinion mining...)
- integrating and executing already existing tools and services

Like the Dicode project, the Dicode workbench has been built using only open source technologies. For getting started with the use of the platform, see this video in AVI or MP4 format.

Recommended Services for gcalle

Service Name	Description	Published by	Date
Similarity Learning - Training	Step 1 of Similarity Learning process	alice	2013-10-0
Entity Prominence Graphs	Displays top entities and distribution over time of entities in a large news corpus.	max,jakob	2012-09-1
demoPinta	Pinta for demonstration at the review	gcalle	2013-11-0

Fig. 7.2 Example of a list of recommended services

- *Workspaces.jsp*—this page lists the workspaces that users have created within the Dicode Workbench.
- *Exit*—this option allows users to logout of the Dicode Workbench.
- **Workspace page**. This page constitutes the main working area for end-users. It allows users to use the services integrated in the Dicode Workbench, customize the workspace to adapt it to their needs, and collaborate with other users to carry out different tasks together.

7.2.2 List of Recommended Services

The Dicode Workbench enables users to register and publish services dynamically. In an initial version of it, users were unaware of the new registered and published services unless the publisher of the service directly contacted them. To alleviate this problem, a personalized list of recommended services was implemented. According to this approach, a different list of services is presented to each user. This list is displayed in the welcome page of the Dicode Workbench after a user logs in, as shown in Fig. 7.2

Figure 7.2 shows a recommendation of three services according to different factors such as, for instance, user preferences, popularity of services and publication date of the service. The recommendation algorithm first considers the publication date of the services, giving more relevance to the more recent ones. Then, it considers the popularity of the service, i.e. how many times a service has been used in workspaces. The more a service is used, the more value it gets. Finally, the algorithm takes into account the annotations of services and gives more relevance to services that are similar to those already exploited by the user.

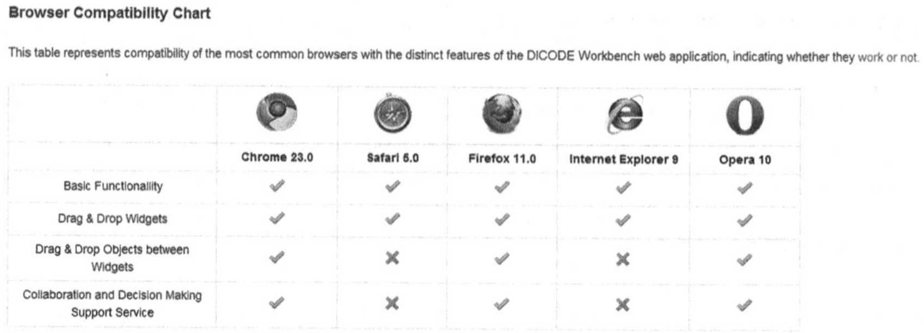

Browser Compatibility Chart

This table represents compatibility of the most common browsers with the distinct features of the DICODE Workbench web application, indicating whether they work or not.

	Chrome 23.0	Safari 6.0	Firefox 11.0	Internet Explorer 9	Opera 10
Basic Functionallity	✔	✔	✔	✔	✔
Drag & Drop Widgets	✔	✔	✔	✔	✔
Drag & Drop Objects between Widgets	✔	✘	✔	✘	✔
Collaboration and Decision Making Support Service	✔	✘	✔	✘	✔

Fig. 7.3 Web browser compatibility chart

7.2.3 Browser Compatibility Chart

In web development, the compatibility of applications with different web browsers constitutes a great challenge because each browser works in a different way. For the Dicode Workbench, we have tried to maximize such compatibility by supporting five major browsers. Unfortunately, it has not been possible to achieve complete compatibility of all features with all of them. To inform users about this compatibility, a compatibility chart has been included in the welcome page, as presented in Fig. 7.3.

7.2.4 Widgets

In the context of the Dicode project, services developed concern scalable data mining, collaboration support and decision making support. Most of these services (at least the ones that users interact with directly) had to be provided in an easy-to-use way, to facilitate their proper use. Towards achieving this, a widget based approach was adopted to deliver the Dicode services to end-users. Widgets in the Dicode Workbench integrate and provide access to all services implemented in the context of the project.

A software widget is a generic type of software application comprising portable code intended for one or more different software platforms. Since the 1980s, when the term "widget" appeared [2], different types of widgets have come up including GUI widgets, disclosure widgets, desktop widgets, widget applications and web widgets. Web widgets, in particular, are software applications/widgets designed for the web and may be embedded and executed within a web page accessed by the end user. They are stand-alone applications and one of their most important features is that the widget host does not control their content. Although the widget content is read-only by the widget host, the end user may interact with the widget, as long as such functionality is provided. On the other hand, the host is able to

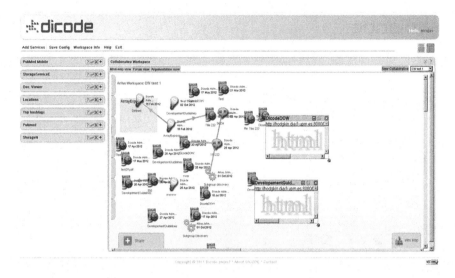

Fig. 7.4 Two column layout of the Dicode workbench

modify the way the widget appears on the host page such as the location or size, and even to establish some initialization parameters in the URI.

A number of Web widget toolkits are nowadays available. Some of the most popular categories of widget toolkits include low level (integrated in the operating system or on top of it as a separate application) and high level widget toolkits (operating system specific or cross platform). Dojo Toolkit [3], jQuery [4], Abstract Window Toolkit [5], Google Web Toolkit [6], YUI, Sencha (formerly Ext JS), TIBCO [7], DHTMLX or Swing [8] are examples of such toolkits.

The widget toolkit selected in Dicode was the Google Web Toolkit (GWT), based on Java and providing a set of core Java APIs and Web widgets. By using GWT, the widget developer may create widgets by writing source code in Java (or by using the GWT Designer provided); the source code produced is, then, compiled to browser "executable" code (JavaScript, HTML, Ajax code) running across all major browsers, including mobile browsers. This is the main benefit of the GWT and the main reason why it was selected. GWT produces different versions of "executable" code, one for each browser, while the developer has to write only one version of code in Java. From a browser's "point of view", each GWT widget ("frame" in GWT terms) is transformed, after the Java code has been compiled, to an HTML "iframe" tag.

One of the main concerns reported by both service developers and end users was the available space for displaying services in the Dicode Workbench. To

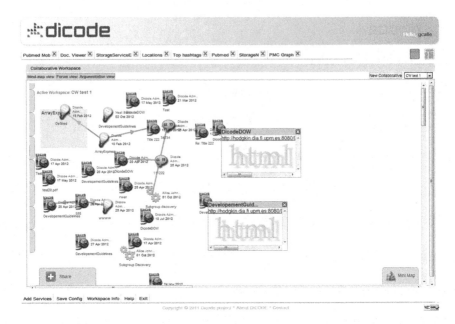

Fig. 7.5 Single column layout of the Dicode workbench

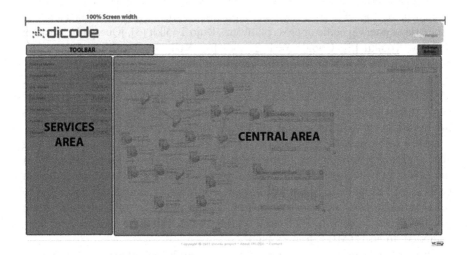

Fig. 7.6 Workspace distribution in the two-column layout

provide an intuitive and usable interface to end users, the working space was designed to support two different layouts: a two-column layout with one column

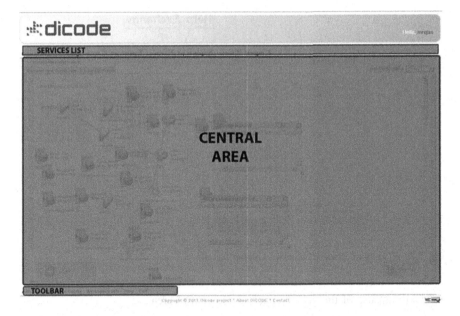

Fig. 7.7 Workspace distribution in the single-column layout

aimed to minimized services and the other column for a maximized service (see Fig. 7.4), and a single-column layout where only one service is displayed at a time (see Fig. 7.5). The Dicode Workbench enables users to easily switch between these two layouts.

The two-column layout suits better to users who use various services and need to exchange and share information between them. The main features of this layout are (see Fig. 7.6):

- 100 % screen width use;
- A left column containing minimized services and a central column for the maximized service;
- Two icons at the upper right corner for enabling users to switch between layouts;
- Collapsible services in the left column for an efficient use of the space;
- Toolbar on top.

The single-column layout suits better to users who intend to work with services handling large amounts of data and need as much space as possible for visualization purposes. The main features of this layout are (see Fig. 7.7):

- 100 % screen width use;
- Services are not presented as widgets but as a list on top of the page;
- Toolbar at the bottom;

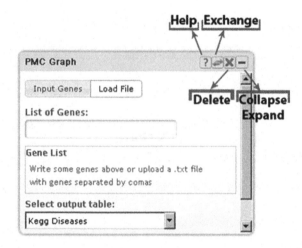

Fig. 7.8 Widgets interface for the two-column layout

- Maximum use of the screen space in the central area.

An example of the appearance of widgets is shown in Fig. 7.8. Widgets provide users with different functions. This interface is only valid in the two-column layout.

There are four buttons located on the upper-right corner of the widgets. Its functionality is:

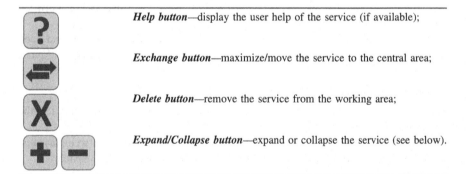

Help button—display the user help of the service (if available);

Exchange button—maximize/move the service to the central area;

Delete button—remove the service from the working area;

Expand/Collapse button—expand or collapse the service (see below).

In the two-column layout, services located on the left column (services area) can be expanded and collapsed by users by clicking on the abovementioned buttons. When the widget is collapsed, the body of the service is hidden and the user can only see the title bar. Figure 7.9 illustrates both these states.

By clicking on the "Help button", a floating window containing help information about the service is presented to the user. The publisher of the service can dynamically customize the information displayed on the floating window. If the publisher does not configure the user help, the system displays an error message

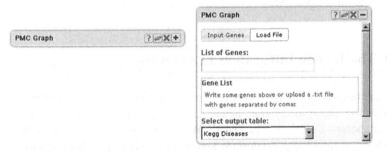

Fig. 7.9 States of the widgets: collapsed (*left*) and expanded (*right*)

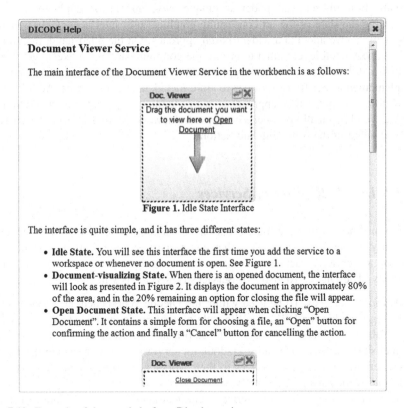

Fig. 7.10 Example of the user help for a Dicode service

indicating that there is not any help available. Figure 7.10 shows an example of the help pop-up window of a service. The help button is available in both layouts.

To enable the publisher of a service to personalize the user help, a field is included into the service publication form called "Help URL". This field enables publishers to indicate a URL where the help of the service is available. This help should be in HTML format.

7.3 The Dicode Integration Framework

As mentioned above, the Dicode Workbench has been implemented as a web application aimed to integrate heterogeneous services, from data mining services to collaboration services. The Dicode Workbench integrates the different services under a common graphic interface and allows users to use them together.

This integration has been performed at two levels: (i) at the user interface (UI) level, and (ii) at the operational level. Integration at the UI level deals with the visualization of services under a unique environment or application. After reviewing the state-of-the-art concerning graphical user interfaces (GUIs) on web applications, we adopted a widget-based approach. On the other hand, integration at the operational level aimed to ensure the communication and exchange of data between different services and applications. For this purpose, we designed and implemented a registry of services to store metadata about services. Additionally, two different integration approaches were defined to permit the integration of services and applications under a common framework. More details on these issues are provided in the following subsections.

7.3.1 Dicode Registry of Services

To ensure access and facilitate location of Dicode services, we implemented a Registry of Services. The idea is that this registry stores information (metadata) related to services, such as service name, service provider, location and functionality. The Dicode Registry of Services (DRS) deals with:

- *Available services.* To be used within the Dicode Workbench, all services must be registered in the DRS. The lifecycle of services in the Dicode Workbench is as follows: (i) service providers develop a new service; (ii) service providers publish the service; (iii) end-users look for services according to their needs, and (iv) end-users add new services to their workspaces to be collaboratively used by the workspace participants.
- *Semantics.* Services are semantically annotated with concepts contained in the Dicode ONtology (DON). Such annotations are stored and managed by DRS. During the publication of services, service providers are asked to annotate manually the new services with concepts retrieved from DON. Such concepts are presented to service providers in the same publication web form, allowing them to select all concepts needed. Services can be annotated according to their functionality, domain, inputs and outputs.

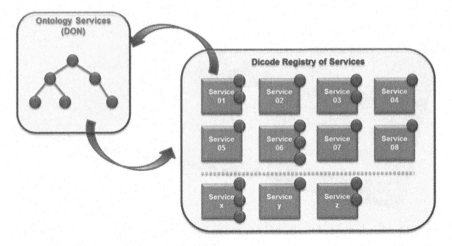

Fig. 7.11 Representation of the Dicode Registry of Services

- *Monitoring information.* This information could be used to evaluate some metrics of services (times used, comments of users, successful attempts, average/min/max times of execution, reliability, etc.).

Figure 7.11 presents the logical view of DRS. As shown, DRS stores the information about the services together with some extra information (metadata) retrieved from DON. In this figure, concepts from DON are represented as green circles. One service can be annotated with none, one or several concepts at the same time. The annotations of services are used by the recommendation algorithm described in Sect. 7.2.2 of this chapter.

7.3.2 Integration Approaches

As stated above, the Dicode Workbench has been designed as a mashup web application, allowing users to share resources under a common framework. Mashup applications usually consist of applications showing together different applications or components such as, for instance, iGoogle. But traditionally, those components neither share information nor communicate between them at all.

In the Dicode project, we moved one step ahead. Users are enabled to move data from one widget to another just using the mouse and the drag-and-drop functionality developed for this purpose. The system architecture is designed to maintain as loose coupling among all integrated resources as possible.

For the design of the integration framework, we considered: (i) the commonalities and differences among the project's use cases to create a general framework with broader applicability, where use cases' scenarios are instances of such a

Fig. 7.12 Main features of the Dicode Integration Framework

framework, and (ii) the integration of existing (3rd party) tools. Finally, we designed an integration framework considering three major issues: flexibility, scalability and sustainability (Fig. 7.12).

Flexibility is needed to integrate any kind of service or application both internal and external to Dicode project. This framework should also consider both existing and future resources. **Scalability** is also a desirable feature to be able to expand the system over time. In the Dicode Integration Framework, scalability is automatically given by the widget approach selected for the development of the Workbench. The widget-based approach allows service providers maintaining their services running on their own machines. Thus, no big servers are needed to deploy the platform even if a large number of resource-demanding services are integrated. Services remain distributed and the computational load is also distributed. Finally, **sustainability** is an essential feature mainly after the end of the project. The framework should allow integrating new services in an easy way, even by the end-users, without any code modification.

To deal with these issues, two different approaches have been designed and adopted in Dicode to integrate services and applications:

- *Light integration*. It can also be called "visual integration". It concerns the traditional mash-up approach, i.e. services/applications are displayed together within the same web interface. No interactions happen between services, thus each service works as a standalone application.
- *Full integration*. Not only services are displayed within the same framework but data can be exchanged between them. Different mechanisms have been developed to communicate data among services. Web interfaces of services need to implement a set of functions to properly carry out the communication.

The Dicode Workbench implements both integration methods. Service developers can select the level of integration desired for their services. More details of these two approaches are provided below.

7.3.2.1 Light Integration

This is the strategy followed by the mash-up approach where different components are just displayed together within a common interface. To carry out this integration, service developers only need to develop the service, develop the web interface, deploy both elements in a web server and publish the service (URI to the web interface) in the Dicode Workbench. After these steps are successfully completed, services can be located and added by users to their shared workspaces.

7.3.2.2 Full Integration

This is the integration approach that allows interactions between components integrated within the same common platform at both the graphical (concerns end-users) and the programmatic (concerns developers) level. Interactions in the Dicode Workbench are events triggered when users move (drag with the mouse) items from one widget to another. Another extra feature of full integration is the definition of mechanisms to preserve state of services between invocations. Obviously, full integration covers all features of light integration.

We designed a loosely coupled architecture based on the idea of message passing interfaces following a publish-subscribe design pattern. In particular, we focused on the *postMessage* mechanism provided by HTML5. This mechanism allows applications running in different windows to communicate information (plain text) across different origins and domains. Although the content of the message can only be plain text, this is enough to communicate almost everything using, for instance, URIs or REST references.

The Dicode Workbench acts as a message mediator between the different services or widgets. When the Dicode Workbench detects that the user wants to move one element from one widget to another, it takes the reference from the origin source and sends a message containing the reference to the target widget. Then, the target widget receives the message, interprets it and performs the actions associated with it. Both reception and sending of messages are optional for widgets (iframes), and it lies in the responsibility of service developers to incorporate them. Additional technical details for service developers are given in the next sections.

7.4   Developing Services for the Dicode Workbench

The Dicode Workbench has been designed and implemented as a web application based on widgets. Depending on the layout, widgets are distributed in one or two "logical" columns. In the two columns layout, minimized widgets are displayed on the left column and, by default, the collaborative workspace is maximized in the center. The Dicode Workbench allows users to maximize any of the widgets located on the left column. When a widget is maximized, it swaps its position with

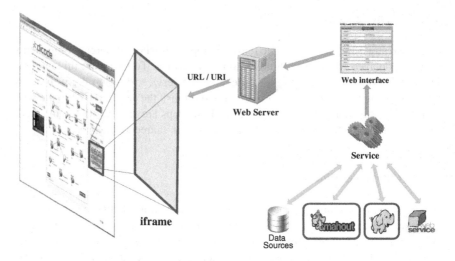

Fig. 7.13 Structure of a service integrated within the Dicode workbench

the widget at the center. Users sharing a workspace are always presented with the same set of services. In any case, users can customize their own view of the shared workspace by rearranging the widgets. The Dicode Workbench allows users to search and add services that have been registered in the system by the service providers.

The only technical requirement for a service to be integrated into the Workbench is that it can be loaded and displayed into an *iframe*. Almost any web application can be displayed inside an iframe. The recommendation for service developers was to use state-of-the-art web technologies such as HTML5, CSS3, JavaScript or jQuery.

Figure 7.13 depicts the structure of a service integrated within the Dicode Workbench. A service performs a concrete task or set of tasks, for instance, retrieves information from a database, analyzes datasets or executes complex algorithms. The results of this task are presented through a web interface. This web interface is deployed in a web server and is accessible via a URL/URI. The Dicode Workbench uses this URL/URI to display the web interface within an iframe element. In this way, a bi-directional communication can be established, i.e. information may flow from the service to the Dicode Workbench and from the Workbench to the service.

To integrate an application or a service in the Dicode Workbench, service providers have to consider the following:

- **Develop the service itself**, i.e. to implement the "logic" of the service. A service might be as simple as displaying a message or perform an addition, or as complicated as running complex algorithms using high performance toolkits. These services can use any technology or library, independent of the Dicode Workbench. Developers should also take into account that the service may be invoked from another application. Thus, a public application interface should be created. We recommend to create web service interfaces based on RESTful services or WS-* (SOAP) services [9].

- **Develop a web interface** to provide users with access to the service. Users have to be able to invoke and use the service from this web interface. In fact, this web interface acts as a wrapper for the service. Additionally, if the service needs or can be invoked with different parameters, the web interface could also provide facilities for the user to input such parameters. Designers and developers should consider carefully the available space devoted to widgets in the Dicode Workbench. Applications for widgets are more similar to mobile applications which have a limited display area. In particular, for the Dicode Workbench, widgets can have two possible states: maximized in the middle or minimized on the sides. Ideally, the web interface should adopt a liquid and elastic design [10].
- **Deploy the service and the web interface**. Both elements have to be accessible through the Internet via an URL/URI. Thus, they have to be deployed in a (web) server.
- **Register or publish the service.** The Dicode Workbench can only display services that have been previously registered in the system. As stated above, the Dicode Workbench maintains a registry of annotated services.

7.4.1 Registration/Publication of Services

In the Dicode Workbench, services can be published at any time by any user of the system. We tried to maintain the publication process as simple as possible; thus, only short information about services is required. As shown in Fig. 7.14, registration of services can be done by selecting the option **Services** from the menu on the left (1) and then, clicking on the link + **Publish a new service...** (2).

Fields required to publish a new service within the Dicode Workbench are:

- **Name**: A unique textual string given by the publisher or service provider to identify the service within the Dicode Workbench.
- **Alias**: A short textual identifier of the service. It is used to display the name of the service on the top of the widget. It is limited to 15 characters.
- **Type**: This attribute allows annotating the service according to the type of service. The values permitted are: "acquisition", "processing" and "visualization".
- **Description**: Free textual description of the service. The publisher or service provider can use this field to provide information about the service(e.g. functionality, parameters required, domain, etc.).
- **Sensemaking operations**: This field is used to annotate services according to the sensemaking operations contained in the Dicode ONtology (DON). None, one or several options can be selected. Annotations are used by the Dicode Workbench to facilitate users' search when they are looking for new services to be added to their workspaces and for the process of recommending services.

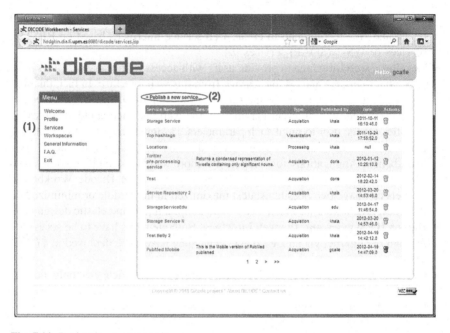

Fig. 7.14 Registering a new service

- **URI**: This is the most important field regarding a service. It establishes the URI where the service is running. The Dicode Workbench uses this URI as iframe source in the widget. Only services accessible via URI can be integrated within the Dicode Workbench. The URI must contain the complete address (with parameters if needed) to invoke the service. Therefore, the service has to be previously deployed in a web server.

7.4.2 Authentication of Services

The Dicode Workbench performs user authentication when users login into the system. In case that a service requires an extra verification due to security restrictions or policy in the organization of the service provider, invocations can be filtered by the IP address of the requester. In Dicode, invocations will be done from the server where the Workbench is deployed. For the Dicode project, the server name was "hodgkin.dia.fi.upm.es", and the IP address was 138.100.11.177. This IP address should be added to the trusted IPs in the server or firewall of the service provider.

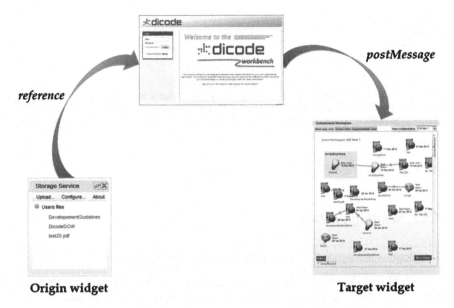

reference

postMessage

Origin widget   **Target widget**

Fig. 7.15 Communication between widgets in the Dicode workbench

7.5 Integrating a Service Within the Dicode Workbench

As discussed in detail in Sect. 7.3.2, the Dicode Workbench offers two integration modes: light integration, which follows a classical mash-up approach, and full integration, which allows communication between widgets. Full integration relies on the HTML5 *postMessage* mechanism, which allows applications running in different windows to communicate information (plain text) across different origins and domains. In the Dicode Workbench, this mechanism is triggered by drag-and-drop actions. Specifically, when the Dicode Workbench detects that the user moves an element from one widget to another, it takes the reference from the origin source and sends a message containing the reference to the target widget (Fig. 7.15. Upon receiving the message, the target widget interprets it and performs the associated actions. In the following sections, we provide some useful instructions to developers for incorporating these functionalities. These instructions contain examples implemented in JavaScript using the facilities of HTML5. Methods proposed can be refined by service developers by using, for instance, jQuery or other libraries.

At the moment, interactions between widgets are triggered by users when they move elements from one widget to another. However, this architecture could be extended to allow widgets/services to trigger events and send data to other widgets/services by following a publish-subscribe design pattern.

```
<!doctype html>
<html>
    <head>
        <script src="js/dragDrop.js"></script>
    </head>
    <body>
        <a href="#" draggable="true" ondragstart="processDragStart(this.href, event)"> item1 </a>
        <a href="#" draggable="true" ondragstart="processDragStart(this.href, event)"> item2 </a>
        <a href="#" draggable="true" ondragstart="processDragStart(this.href, event)"> item3 </a>
    </body>
</html>
```

Fig. 7.16 Example code to allow items to be dragged

7.5.1 Sending an Item

Two actions are needed to send items in the Dicode Workbench:

- All HTML elements which can be dragged must be labeled as *draggable*;
- The information to be sent when the item is dragged must be defined.

To label an element as draggable, an attribute to the tag of this element has to be added. In such a way, browsers can identify those elements as draggable. In the code of Fig. 7.16, a file named *dragDrop.js* containing JavaScript code is imported. This file contains the functions to handle the drag behavior of an item and to receive messages. The full content of *dragDrop.js* file is shown in Fig. 7.17.

To establish the information to be sent, one needs to add an event to the draggable items and define a function to process the event. Usually, this function will define the information to be exchanged between widgets. There are two options to add an event to an item:

- Including the event in the HTML code as shown in Fig. 7.17. `ondragstart` is the name of the event that is triggered when users start a drag action; `processDragStart` is the name of the function to process the event.
- Invoke the function `addEvent`, included in the code of Fig. 7.17, for each draggable item. Using tools and libraries such as, for instance, cssQuery, the DOM of the document can be examined looking for items defined as draggables.

Once the listeners for the events are established, service developers have to codify the function(s) to attend the event(s). As stated above, in the Dicode project we adopted a message passing strategy. Thus, this function should be used to construct the message that the service wants to communicate to the other services. For the needs of the project, a preliminary set of messages was defined (see Sect. 7.5.3). An example of such a function is provided in Fig. 7.17. In this case, one message is created following the message structure adopted. After the message is created, it is sent to the parent window (i.e. the Dicode Workbench).

```
/* dragDrop.js */

var addEvent = function(obj, evType, fn) {
                    if (obj.addEventListener) { //W3C DOM
                        obj.addEventListener(evType, fn, false);
                    } else if (obj.attachEvent) { //IE DOM
                        obj['e' + evType + fn] = fn;
                        obj[evType + fn] = function() {
                                            obj["e" + evType + fn](self.event);
                                        };
                        obj.attachEvent("on" + evType, obj[evType + fn]);
                    }
                };

function processDragStart(ref, e) {
                var message;
                message = '["Item": {\n' +
                            '\t"type": "File",\n' +
                            '\t"name": "",\n' +
                            '\t"uri": "'+ref+'",\n' +
                            '\t"format": "",\n' +
                            '\t"description": ""\n' +
                            '\t}\n}';
                if (parent.postMessage) {
                    parent.postMessage(message,"*");
                } else {
                        alert("Your browser does not support the postMessage");
                }
            };

function OnMessage (event) {
        var message = event.data;
        //Check the location of the caller

        //Opera earlier than version 10
        if ('domain' in event) {
            if (event.domain != "hodgkin.dia.fi.upm.es:8080") {
                return;
            }
        }

        // Firefox, Safari, Google Chrome, Internet Explorer from version 8 and Opera from version 10
        if ('origin' in event) {
            if (event.origin != "http://hodgkin.dia.fi.upm.es:8080") {
                return;
            }
        }
        // TODO Treatment of the received message
        alert("RECEIVED: "+message);
    };

onload = function () {
                    addEvent(window, "message", OnMessage);
                };
```

Fig. 7.17 Complete code of JavaScript file *dragDrop.js*

Type of item	Message format
| File | ```
{Item: {
        type: File,
        name: name,
        uri: uri,
        format: format,
        description: description
    }
}
``` |
| Image | ```
{Item: {
        type:  Image,
        name: name,
        uri: uri,
        description: description
    }
}
``` |
| Text | ```
{Item: {
        type:  Text,
        content: content
    }
}
``` |
| Link | ```
{Item: {
        type:  Link,
        uri: uri
    }
}
``` |

Fig. 7.18 Basic set of message formats

7.5.2 Receiving an Item

To receive messages in an application, two actions have to be carried out:

- Create a listener to receive messages in the application/service. An example of how to create such a listener is shown at the end of Fig. 7.17. This listener will be associated to the window/iframe where the application is running.
- Define and codify the treatment of the information received into the message. In the example of Fig. 7.17, the function OnMessage has been defined for this purpose. In this case, the function OnMessage checks the origin of the message to prevent from unauthorized uses, and then the content of the message is shown in a pop-up window. Treatment of messages can be as simple as presented, but it can be as complex as service developers need.

7.5.3 *Message Formats*

Message passing approach requires that both the sender and the receiver agree on a common format for exchanging information. In the context of Dicode, we adopted a message format based on JSON. A preliminary set of basic message formats has been defined to be used by the applications/services within the Dicode Workbench (see Fig. 7.18—note that parts highlighted in yellow have to be completed by the sender with the proper information). This set of formats is not a closed list; it could be easily expanded with more types of messages.

7.5.4 *Preserving the State*

When a user swaps a "maximized" service with a "minimixed" service, both services need to maintain their current views to ensure continuous usage during a session. To facilitate this task, the Dicode Workbench sends to the services some extra information as HTTP parameters during the invocation call. Services are available through an URL and they are loaded in an iframe element in the Workbench. Two extra parameters are sent to services using the query string:

- expid: it is an integer number identifying the current workspace;
- usrid: it is an integer number identifying the current user (this number is unique for each user in the Dicode Workbench).

Example of invocation:

http://hodgkin.dia.fi.upm.es:8080/StorageService?width=279&height=215&color=0066bb &expid=3&srvid=1&usrid=2"></iframe>

Services should properly process these parameters to carry out the necessary action to preserve the state of services between invocations.

7.6 Conclusions

This chapter has presented work carried out in the context of the Dicode project to handle various technical integration issues. This work focused on two major outcomes: the Dicode Workbench and the Dicode Integration Framework. The former is a web application designed to allow users to access multiple services and applications under a common web interface. The latter constitutes the framework created to support the integration of heterogeneous services.

To support the integration of services, two different approaches were followed: "light integration" and "full integration". Some Dicode services have already implemented both approaches; however, most of them are integrated using the first one. There are two main reasons for that: first, light integration requires fewer efforts from service developers; second, most of the services required by end users do not need to interact and interoperate with other Dicode services. For the majority of end users, it is enough to be able to use different services within a common framework.

In total, more than 25 services have been developed by technical partners during the project to provide end-users (use case partners) with innovative solutions to their problems. All these services have been integrated in the Dicode Workbench. The generic and flexible approach followed in the design of the Dicode Workbench enables its usage in different scenarios and multiple domains (i.e. beyond those elaborated in the Dicode project).

References

1. Hall, M.: More Servlets and JavaServer Pages. Prentice Hall PTR, Englewood Cliffs (2001)
2. Swick, R.R., Ackerman, M.S.: The X Toolkit: More bricks for building user interfaces, or widgets for hire. In: Proceedings of the Usenix Winter 1988 Conference. pp. 221–228. http://s.niallkennedy.com/papers/xtk.pdf (1988)
3. Russell, M.: Dojo the Definitive Guide, 1st edn. O'Reilly Media, Sebastopol (2008)
4. Lindley, C.: jQuery Cookbook. O'Reilly Media, Sebastopol (2010)
5. Zukowski, J.: Java AWT Reference. O'Reilly Media, Sebastopol (1997)
6. Hanson, R., Tacy, A.: GWT in Action: Easy Ajax with the Google Web Toolkit. Manning Publications Co., Greenwich (2007)
7. Brown, P.C.: TIBCO Architecture Fundamentals. Addison-Wesley Professional, Upper Saddle River (2011)
8. Eckstein, R., Loy, M., Wood, D.: Java Swing. O'Reilly & Associates, Inc., Sebastopol (1998). Google Web Toolkit, http://code.google.com/intl/el-GR/webtoolkit/
9. Pautasso, C., Zimmermann, O., Leymann, F.: Restful web services vs. "big" web services: making the right architectural decision. In: 17th International Conference on World Wide Web, 21–25 April 2008, Beijing, China (2008)
10. Cederhlm, D.: Bulletproof Web Design: Improving Flexibility and Protecting Against Worst-Case Scenarios with XHTML and CSS. New Riders, Berkely (2006)

Chapter 8
Clinico-Genomic Research Assimilator: A Dicode Use Case

Georgia Tsiliki and Sophia Kossida

Abstract Biomedical research becomes increasingly interdisciplinary and collaborative in nature. Researchers need to effectively collaborate and make decisions by meaningfully assembling, mining and analyzing available large-scale volumes of complex multi-faceted data residing in different sources. Through a real scenario, this chapter reports on the practical use of the Dicode solution in the above context. Evaluation results show that the proposed solution enables a meaningful aggregation and analysis of large-scale data in complex biomedical research settings. Moreover, it allows for new working practices that turn the problem of information overload and cognitive complexity into the benefit of knowledge discovery.

Keywords Genomics · Transcriptomics · Gene ontology · Integration

8.1 Introduction

The field of biomedical research has recently seen a vast growth in publicly available biomedical resources, including multiple types of datasets and databases. A major advance is that now researchers have access to complementary views of a single organism by analyzing multiple types of data, including whole genome sequencing, expression profiling and other high-throughput experiments. These data, which are often called '-omics' data, include the genome sequencing data (genomics), the complete set of RNA transcripts produced by the genome and

G. Tsiliki (✉) · S. Kossida
Biomedical Research Foundation, Academy of Athens, Athens, Greece
e-mail: gtsiliki@bioacademy.gr

S. Kossida
e-mail: skossida@bioacademy.gr

N. Karacapilidis (ed.), *Mastering Data-Intensive Collaboration and Decision Making,* 165
Studies in Big Data 5, DOI: 10.1007/978-3-319-02612-1_8,
© Springer International Publishing Switzerland 2014

analysed via microarray, real-time PCR or Next-Generation Sequencing platforms (transcriptomics), protein structures and function (proteomics), or any other data available for the organism under study, and provide novel views of cellular components in the biological systems [1]. As a consequence, an enormous amount of digital content is produced everyday (i.e. information that is created, captured, or replicated in digital form as well as hundreds of analysis systems), resulting in high rates of new information being distributed and demanding attention [2].

Most of those datasets are well organised in publicly available databases, although there are existing limitations in accessing, storing, mapping and managing the increasing amount of data available [3]. Managing the amount and diversity of -omics data is a task that must be supported by appropriate algorithmic analysis and software tools [4]. Moreover, there have been attempts to algorithmically unify the above mentioned data [5] and their supplementary views [6]. However, choosing the right datasets, databases and tools for a given project is difficult even for an expert, which increases the importance of handling the data in a scientifically sound way [7]. To address such limitations, cloud and distributed computing, schema-free solutions, domain-specific and process-oriented programming languages or special algorithmic solutions are applied [7–9].

Overall, there is a growing need for data and computing resources to be readily reused, repurposed and extended by other scientists [9]. For instance, the well-known Galaxy Project (http://galaxy.psu.edu/) offers a web-based platform allowing researchers to perform and share complete analyses; the GenePattern platform (http://www.broadinstitute.org/cancer/software/genepattern/) provides access to more than 180 tools for genomic analysis to enable reproducible in silico research. Other attempts, such as BioMart (http://www.biomart.org/) or its cancer specialized version IntOGen (http://www.intogen.org/home), focus on linking biological databases [10, 11].

Within this environment, biomedical research has become increasingly interdisciplinary and collaborative in nature [9, 12]. The increasingly specialized resources show that the way forward is to form biomedical research collaboration teams to address complex research questions. Such interdisciplinary teams would better meet challenges relative to various problems such as how to store, access, analyze and integrate multiple types of data [7]; or, how to work with multiple databases simultaneously [13]; or even, how to make data accessible and usable to life sciences researchers [3]. In addition, tools facilitating sense- and decision-making by appropriately capturing the collective intelligence that emerges during such collaboration are lacking. Biomedical researchers need such tools to efficiently and effectively collaborate and make decisions by appropriately assembling and analyzing enormous volumes of complex multi-faceted data residing in different sources. Supporting team collaboration under such circumstances is still considered a challenging task [14].

Dealing with these issues, Dicode's *Clinico-Genomics Research Assimilator (CGRA) Use Case* aims to support the entire life cycle of biomedical collaboration by supporting clinical researchers and bio-scientists allowing them to easily examine, reuse and interpret heterogeneous clinico-genomic data and information sources, as well as to reach decisions for the production of new insightful conclusions, without having to worry about the method of locating and assembling huge quantities of data. Towards the accomplishment of this goal, a number of Dicode services were developed.

8.2 The CGRA Use Case

CGRA is built as a general research assimilator environment able to handle large data, such as Next Generation Sequencing (NGS) data. The Dicode services developed for CGRA aim to underline interdisciplinary collaboration and decision-making by facilitating integration of available information under a common platform in the following ways:

- identify data and annotation databases—data repositories and annotation data to augment the data size and relative information;
- social sharing—exchange valuable experiences and tacit knowledge;
- data sharing—exchange 'similar' or important data for augmenting data size;
- manipulate large datasets—provide means to efficiently handle large amounts of data and avoid 'out-of-memory' errors;
- tacit knowledge—provide suitable means to establish and record expert's knowledge and experience through a trustworthy evaluation framework;
- analysis practices/tools—building and sharing predictive models for data analysis, as well as integration models or templates for merging heterogeneous data sources;
- efficient use of expertise and results—provide support in decision-making, if possible, throughout the entire biomedical task in a collaborative setting;
- efficient presentation and grouping of tools/results—suggest ranking of tools, services or users based on the relevancy of the issue at hand;
- biomedical resources sharing and interconnection—in summary, provide a social collaborative network for interconnection and interaction between biomedical researchers, tools, data and resources.

8.2.1 CGRA Everyday Practices

CGRA is planned and materialized to seamlessly link and mine disparate clinico-genomic data sources, and meaningfully support the whole life cycle of a biomedical experiment. The breast-cancer case was initially targeted; nevertheless, other diseases (e.g. cardiovascular disease) or organisms (e.g. plant data) can be considered.

Biomedical researchers often augment their in-house data with publicly available data stored in varying formats (see Table 8.1). A typical process is to download the raw or pre-processed data from a database (e.g. GEO) along with all

Table 8.1 Input data considered for CGRA. Note that GEO is an NCBI repository which is separately presented here since it is one of the main CGRA's data sources

Data type	Data description	Web sites	Data in numbers	Data in size
Omics data	Gene-expression data (normalized or raw data)	Gene Expression Omnibus (GEO)	3,413 datasets and 1,032,797 samples	~ 500 Kb/sample
		ArrayExpress	44,245 experiments and 1,253,392 assays	~ 32 Mb/dataset
		Stanford Microarray Database (SMD)	82,542 experiments	
Phenotypic data	Supplementary, clinical or phenotypic data available	Same as 'Omics data' or individually supplied files	on average2 files per dataset	~10 Kb/dataset
Molecular Pathways (MP)	Data from known and established molecular networks	Kyoto Encyclopedia of Genes and Genomes (KEGG)	416 pathway maps (153,758 total)	
		Reactome	3,931,211 data entries	
Annotation data	Reference databases for biomedical and genomic information	Gene Ontology (GO)	~30,000 terms	
			~50,000 relationships	
		National Center of Biotechnology Information (NCBI):	1. 26,473 annotated coding regions	
		1. RefSeq	2. 129,493 homo sapiens entries	
		2. UniGene	3. ~126,551,501,141 bases	
		3. GenBank	4. >21 million citations for biomedical literature	
		4. PubMed		

the relevant phenotypical and clinical information needed to understand and analyze the data. An intermediate but important step is to reformat and store them locally, in order to visualize and analyze them. The analysis could be conducted by using either a standalone tool, such as Cytoscape (www.cytoscape.org), or in-house scripting using, for example, the R statistical software environment (www.r-project.org).

Perhaps the most important step in the life cycle of an experiment is to interpret and communicate the findings. The results need to be comparatively assessed against modern methodologies, but most importantly they need to be biological or medically interpreted to have an insight into the initial biological question of interest. For that purpose, researchers confer with databases, such as KEGG, or standalone tools which are directly linked to databases and can qualitatively and quantitatively assess the submitted results using the database resources. Decision-making plays an important role in this context; scientists need to evaluate their options of analyses, databases, tools, and often base their decisions on past experience and feedback from their colleagues. The CGRA use case is founded on an integrated knowledge discovery scenario that amalgamates and maps gene-expression profiles onto gene-regulatory networks in an aim to uncover molecular regulatory mechanisms that govern target phenotypes.

8.2.2 CGRA Related Dicode Services

CGRA aims to integrate information and knowledge in a clinico-genomics setting. The *Dicode Workbench* (see Chap. 7) provides the means to access and critically assess the essential resources and tools in a common interface which bundles all functionalities together. It is the integration platform for all Dicode data analysis and support services. The *Storage service*, built for all Dicode use cases to comfort the sharing and exchange of information (files, reports, etc.) in data-intensive and cognitively-complex settings, is embedded within the Workbench. This service provides all functionalities needed to allow permanent and reliable storing of files as well as their accessibility.

Other Dicode services exploited in CGRA are:

- *Collaboration and Decision Making Support services* (see Chap. 6). The Collaboration Support services exploit the reasoning abilities of humans to facilitate sense-making of the Dicode data mining services' results and capitalize on their outcomes. The *Decision Making Support services* translate the information and knowledge available into machine interpretable data in order to allow active participation of the system in collaborative activities.
- *Forum Summarization service.* This service receives clusters of discussion threads as input from relevant public forums and identifies their most prominent terms (topics). The identified topics can be used to derive the main theme in the cluster supplied.

- *Subgroup Discovery service* (SD, see Chap. 5). This service searches for subgroups in any user provided data by searching the rules that cover target and non-target value examples [15]. Particularly, SD finds patterns that describe subsets of a dataset that are highly correlated relative to a target attribute. It supports two different SD data mining algorithms.
- *Recommendation service* (see Chap. 5). This service recommends similar users or documents from log file data based on similarity models learned by using the Dicode Similarity Learning Service. Specifically for CGRA, the GEO-Recommender (GEOR) web-based application is employed to search the GEO database for appropriate datasets based on keywords or the description supplied by the user.
- *PubMed service*. It searches for relevant (to the topic of discussion) scientific articles from the PubMed database (http://www.ncbi.nlm.nih.gov/pubmed).

Overall, the Dicode services exploited in CGRA:

- facilitate and augment the collaboration and decision-making in data-intensive and cognitively-complex settings;
- facilitate the sharing of knowledge on data, analysis methodologies or web tools;
- apply methodologies to derive predictive disease models (diagnostic and prognostic) such as feature selection and classification analysis;
- utilize mining methodologies in the context of clinico-genomic research field.

8.3 A Real Scenario

To better illustrate the use of the proposed Dicode solutions, this section presents an illustrative scenario concerning collaboration in the area of breast cancer research.

Alice is a Pharmacology PhD student. Her research is on adjuvant hormonal therapy for patients with breast cancer disease; particularly, she is interested in identifying how Tamoxifen (Tam) resistant cells modulate global gene expression. Tam is a widely used antagonist of the estrogen receptor (ER), whereas its resistance is a well-known obstacle to successful breast cancer treatment [16]. While adjuvant therapy with Tam has been shown to significantly decrease the rate of disease recurrence and mortality, recurrent disease occurs in one third of patients treated with Tam within 5 years of therapy. Alice selected and analyzed gene-expression data from 300 patient samples with the help of Neal, an MD at a collaborating university hospital, and Jim, a postdoctoral researcher in Bioinformatics. These data are derived from whole human genome expression arrays (Affy U133A Plus 2.0—see http://www.affymetrix.com). Although the sample is relatively large, Alice believes that augmenting the data with publicly available data will be a good idea for statistically significant results.

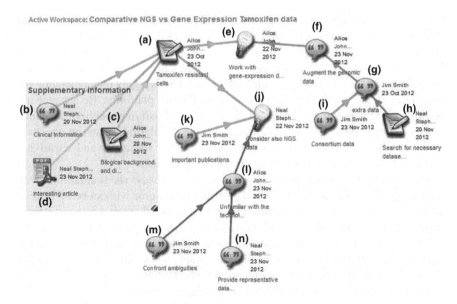

Fig. 8.1 Launching a collaboration workspace for estimating the dominance of Tamoxifen resistance cells to global gene expression. Alice and her colleagues upload and link related biological, clinical or technical information

To analyze the data and discuss the analysis results, Alice, Neal and Jim decide to collaborate by using the Dicode Workbench and, specifically, the mind-map view of Dicode's *Collaboration Service* (see Chap. 6). Alice is launching a new collaborative workspace (Fig. 8.1). Even though all three collaborators are aware of the benefits and difficulties of Tam treatment, Alice adds a note on the collaboration workspace to fully explain the characteristics of the genomic data (Fig. 8.1a). Neal has collected all the necessary clinical information and posts them on the collaboration space (Fig. 8.1b). Apart from stating the scientific question of interest, Alice summarizes the biological background and technical difficulties around it (Fig. 8.1c), while Neal finds an interesting article concerning the Tam treatment and uploads the corresponding pdf file on the workspace (Fig. 8.1d). In the mind-map view, users may group together related items by using colored rectangles (see, for instance, the one entitled "Supplementary information", drawn by Neal).

Alice believes that they should first work with the gene-expression data (idea item (e), Fig. 8.1) and, moreover, they should augment the genomic data (comment item (f), Fig. 8.1). Jim suggests launching the GEOR service (Fig. 8.1g) to find "similar" datasets in terms of pathology characteristics. Neal offers to find the extra datasets (Fig. 8.1h), since he is more confident with the technical characteristics of the data. Jim agrees (Fig. 8.1i), and adds that there are data available from consortiums such as caBIG (https://cabig.nci.nih.gov/), which have extensively proven the need to augment or at least compare and assess findings across multiple datasets.

Even though Alice believes that they should first work with the gene-expression data, Neal argues that they should also consider NGS data (idea item (j), Fig. 8.1). He mentions that he is responsible for a clinical trial and can have access to total RNA from human breast cancer cell lines, which are then analyzed using NGS technology. Jim is also working with NGS data and he is highly recommending the integration or at least the comparative study of the two platforms. He has recently published some important results (Fig. 8.1k) by classifying publicly available transcriptomics data and he has found striking similarities between the two. Moreover, NGS is the latest technology having higher specificity and sensitivity, and thus they could meaningfully augment Alice's results.

Alice is reluctant to start working with NGS data because she is unfamiliar with the technology and argues that she will probably invest time without being assured about the significance of the results (Fig. 8.1 (l)—note that arrows in red denote argumentation against the 'parent' item, while arrows in green denote argumentation in favor). To defeat this statement, Neal suggests (Fig. 8.1m) to provide her with a representative dataset from his laboratory, while Jim offers to help her (Fig. 8.1n) deal with all the ambiguities between the two datasets.

Alice considers exploiting the SD data mining algorithm. This is a popular approach for identifying interesting patterns in the data, since it combines a sound statistical methodology with an understandable representation of patterns. For example, in a group of patients that did or did not respond to specific treatment, an interesting subgroup may be that patients who are older than 60 years and do not suffer from high blood pressure, respond much better to the treatment than the average.

To invoke the SD algorithm, Alice uploads the associated service item on the workspace (Fig. 8.2a) and follows the necessary configuration steps to start the execution of the service. Configuration includes the specification of the URI for the REST-based SD service and specification of parameters such as input file, number of rules to be used, service ontology, and minimum number of subgroups to be retrieved. Jim advises her on the SD methodology parameters (Fig. 8.2b); particularly, they decide to run the algorithm with a minimum number of four subgroups for each biological category to emphasize only the highly ranked statistically significant groups of the data. Alice proceeds in entering the discussed parameters that include the input file containing the data, the number of rules to be used, the service ontology, as well as the list of attributes to be included/excluded (Fig. 8.3). Once these parameters are entered, Alice hits the "Go" button to start the execution of the SD service. The execution of the SD service is done by invoking the REST-based API of the SD service. When the SD service starts its execution, the color of the corresponding icon on the workspace changes (to indicate that the service is currently being executed). This allows participants to get informed on the status of execution of such service items, as their execution can be time consuming.

Upon the successful termination of the service's execution, the corresponding icon on the workspace changes again its color to indicate that the execution was finished (Fig. 8.2a). In addition, the service outcomes are automatically

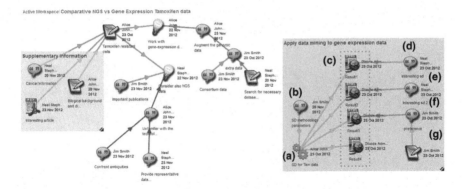

Fig. 8.2 Application of SD service to gene expression data and assessment of results

Fig. 8.3 Form allowing the configuration of the execution of the SD service item

Subgroup Discovery on Gene Data

Select input file:	Browse...
Number of rules:	5
Use ontology:	☑ Biological process ☑ Cellular component ☑ Molecular function
Attributes to include:	
Attributes to exclude:	
Execution:	Go!

꜀dicode

uploaded on the collaboration workspace (Fig. 8.2c). One collaboration item is created for each result (in this scenario, outcomes are in html format). Outcomes are tables with GO and KEGG terms, which describe biological processes related to the estimated groups of genes. For this particular run, the SD results are summarized in the following four subgroups: 'sequence specific DNA binding TFA', 'transcription from RNA polymerase II promoter', 'signaling transducer activity', 'PI3 k-Akt signaling' (Results 1–4, Fig. 8.2c). The findings of the SD service seem convincing to Neal (Fig. 8.2d–e), while Jim expresses his opposition about the third outcome and quotes a part of a scientific paper he recently read (Fig. 8.2f–g).

The same procedure (invoking the SD service and collectively assessing its output) is then followed for the NGS data (Fig. 8.4a–b). The three researchers carefully examine the commonalities between the two SD runs (on genomic and NGS data) and share their insights. The subgroups returned for the NGS data (Fig. 8.4c) are very similar to the ones obtained from SD service on genomic data (Results 1–4 correspond to: 'response to stimulus', 'positive regulation of transcription', 'transcription from RNA polymerase II promoter', 'signaling transducer activity'). Alice is impressed with the commonalities found between the two SD runs; she is now convinced that there is scope to integrate additional NGS data.

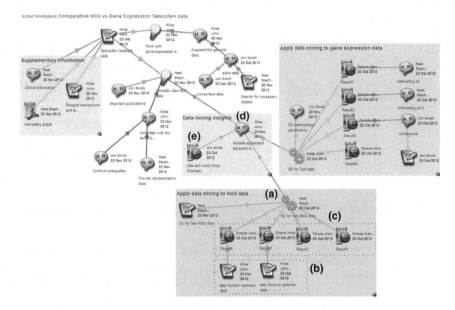

Fig. 8.4 Application of SD service to NGS data, assessment of results, and insights

She expresses her insight (Fig. 8.4d) and links it to the original Neal's idea (note that SD service items are also linked as arguments in favor of this insight). To further elaborate this issue, Jim uses the PubMed service offered through the Dicode Workbench to search for recent relevant literature. He then uploads a link (Fig. 8.4e) pointing to a scientific report that strengthens Alice's argument.

The above collaboration may proceed to further augment the gene expression and NGS data. For instance, as Jim has previously suggested, the researchers involved may invoke GEOR to continue the analysis with the datasets that Neal has already downloaded.

8.4 Evaluation Process and Feedback Received

The evaluation process of the project was conducted in two phases. During the evaluation process, the Dicode Workbench was assessed together with the Collaboration and Decision Making Support services, as well as the SD and Recommender services. Particular emphasis was given to the Collaboration service. For the second evaluation round, a scenario very similar to the one presented in Sect. 3 was distributed to evaluators in order to assess the degree of Dicode's flexibility to serve all members of the collaborating group. In this section, we present a summary of the two phases of the evaluation process, together with statistical results from the evaluators' answers.

8.4.1 First Evaluation Round

During the first evaluation round of CGRA use case, the Dicode Workbench and five Dicode services were evaluated in terms of usability, accessibility and acceptability. At the time of the first evaluation round, two of the services were integrated into the Dicode Workbench, whereas the remaining three were evaluated as standalone software. The Dicode services were evaluated by 61 volunteers from the four participant countries of the project (Greece, Spain, Germany, United Kingdom) who were selected using the *snowball sampling* method [17]. Questionnaires distributed to evaluators aimed at gathering mainly quantitative but also qualitative feedback when appropriate. Evaluators were asked to carefully read the related instructions, have a 'hands-on' session for each service, and finally fill in the questionnaires.

Based on the feedback received from the first evaluation phase, the Dicode Workbench was reported to be a promising tool, which facilitates users to set their research objectives and better understand the data and methodologies used in their research. The vast majority of the evaluators appreciated the potential of exploiting the synergy of machine and human reasoning through data mining and collaborative decision making services. The innovative approaches on the text-mining services seemed to be appreciated by the evaluators, who generally agreed on the usefulness and acceptability of the provided services. Nevertheless, additional work towards the improvement of Dicode services in terms of their documentation, user interface and performance seemed to be essential. Another issue raised relates to testing these services in various data-intensive contexts, in order to further assess their applicability and potential, and gradually build their generic nature.

8.4.2 Second evaluation round

For the second evaluation round, rather than evaluating the usability of each service, we were interested in evaluating how those tools facilitate collaborative processing of the tasks at hand. Thus, the goal of the second evaluation round was to simulate a multi-tasking environment where users need to work on diverse CGRA tasks. Since all services were integrated into the Dicode Workbench with all their foreseen functionalities already implemented, the second evaluation round was conducted by recruiting senior members of the biomedical community. Based on related literature suggestions [18], we combined two evaluation methodologies, namely scenario-formed video-casts and questionnaires, in order to capture experts' judgements about the usage of Dicode services and their overall ratings in an effortless way. Similarly to the approach described in [19], an overview of the suite of services was presented to real expert users through a concrete everyday usage scenario (see Sect. 3), who were then requested to imprint their thoughts about services' functionalities as well as their overall opinion of the Dicode Workbench by providing quantitative and qualitative remarks.

Our goal was to measure the usability, acceptability and functionality of the Dicode Workbench and its integrated Dicode services, by directly facing some important issues the biomedical community needs to cope with, and particularly:

- Data analysis issues, and
- Understanding and assessing data analysis findings based on collaboration and decision-making in a multi-disciplinary environment.

Ten participants were recruited with different areas of expertise, such as Bioinformatics and Biology, in order to cover important and currently evolving sub-areas of biomedical research and bioinformatics. Special care was given to their seniority; two special groups were considered: researchers with 5–10 years and 10–20 years of experience after they acquired their PhD degree. Answers to the quantitative questions of the questionnaires were given for ordinal data in a 1–5 scale (questions concerning the quality, acceptability and accessibility of the services provided), where 1 stands for 'I strongly disagree' and 5 for 'I strongly agree', and for continuous numerical data (scale data) in a 0–10 scale (questions concerning the services' usability), where 0 stands for 'none' and 10 for 'excellent' [20, 21].

Overall, the Dicode Workbench was reported to be intuitive with well integrated services. Evaluators were satisfied with the information provided by the video-cast, although they reported that extra time was needed to familiarize with the Workbench. They were sceptical about adopting new practices but less sceptical for the ability of the Dicode Workbench integrated services to deal with cognitive-complex issues, to enhance collaboration between their peers and in that respect assist exchanging of information and advice. The Dicode collaboration and decision-making support services were highly marked; special mentions include the data management mechanism, and the different manners of sharing or discussing data and results. Furthermore, evaluators reported that the platform offers ease of communication, and strong data/information archiving features. Additionally, we found that the seniority of the evaluators did not affect their responses to the questionnaire. Evaluators deemed that the Dicode Workbench brings potential benefit to their work and provides sufficient services to support their work. Nevertheless, they were reluctant to use the Dicode integrated services. In Figs. 8.5, 8.6 and 8.7, we present the evaluators' responses to the three sets of questions concerning the overall impression from Dicode Workbench, its decision-making options, and its usability, respectively.

Figure 8.5 shows the mode and quartile trend (minimum, median and maximum values) of the responses relative to the evaluators' overall impression of the Dicode Workbench. Specifically, the questions asked were:

- *Question 1: The information provided on the video is clear.*
- *Question 2: The design of the Dicode Workbench is very pleasant.*
- *Question 3: The use of Dicode Workbench is easy.*
- *Question 4: The user interface of Dicode Workbench is intuitive.*

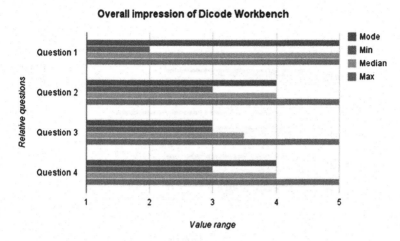

Fig. 8.5 The overall impression of Dicode Workbench: mode, minimum, median, and maximum values are presented for Questions *1–4*. The range of original values is *1–5*, where *1* denotes strong disagreement and *5* strong agreement

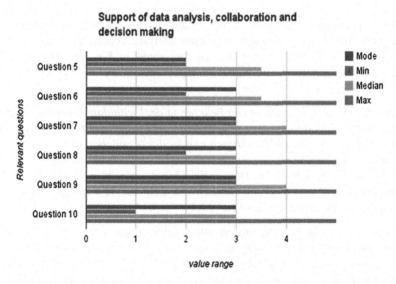

Fig. 8.6 Support of data analysis, collaboration and decision-making: mode, minimum, median, and maximum values are presented for Questions *5–10*. The range of original values is *1–5*, where *1* denotes strong disagreement and *5* strong agreement

Overall mode/median values range from 3 to 5, where the minimum value is 3 except from Question 1 which is 2. Evaluators were satisfied with the information provided by the video (only one value is 2), and its pleasant design (mode and median values equal 3).

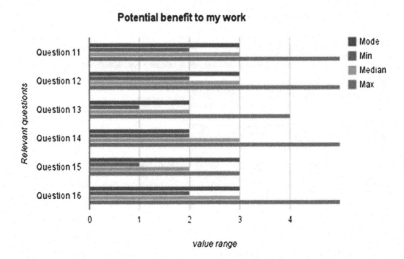

Fig. 8.7 Potential benefit to my work: mode, minimum, median, and maximum values are presented for Questions *11–16*. The range of original values is *1–5*, where *1* denotes strong disagreement and *5* strong agreement

The summarized responses of the evaluators for questions relative to the evaluators' opinion for the support of data analysis, collaboration and decision making via the Dicode Workbench are presented in Fig. 8.6. The questions referred to the usefulness and particular capabilities of the services integrated in the Workbench, i.e.:

- *Question 5: Dicode services can help me to deal with data-intensive issues.*
- *Question 6: Dicode services can help me to deal with cognitive-complex issues.*
- *Question 7: The Dicode Workbench can facilitate collaboration.*
- *Question 8: The Dicode Workbench can enhance decision making.*
- *Question 9: The services of Dicode Workbench are very well integrated.*
- *Question 10: The Dicode Workbench can help me be more productive and concentrate on creative activities.*

As shown in Fig. 8.6, the mode/median values range from 2 to 4, where the mode values are the lowest ranging from 2 to 3. Evaluators were more reluctant towards this second set of questions, which is partly explained by the fact that these questions enquire their impression over scientifically significant matters involving data analysis and decision-making. Researchers are always cautious towards the data analysis methods they use, especially given the complexity of the biomedical data. Another reason of their reluctance should be attributed to the sample distribution, as senior researchers who are confident with their analysis skills are less willing to adopt other analysis routines. Evaluators are less sceptical for the ability of the Dicode Workbench integrated services to deal with cognitive-complex issues, as well as to enhance collaboration between their peers and in that respect help them to be more productive and concentrate on creative activities.

The last set of questions explored the potential benefit of Dicode to evaluators' current work practices, namely:

- *Question 11: I can see the potential benefit of using Dicode Workbench in my work.*
- *Question 12: Dicode provides sufficient services to support my work.*
- *Question 13: I would consider using the Dicode Workbench in the near future.*
- *Question 14: The use of Dicode will have positive impact on my current work practices.*
- *Question 15: The use of Dicode will change my current work practices.*
- *Question 16: I will recommend the Dicode Workbench to my peers/community.*

In Fig. 8.7, we can observe that evaluators are keen to change their current work practices, although there are instances of very low or very high marks for all six questions considered. Overall, evaluators believe that the Dicode Workbench brings potential benefit to their work (mode, median equal 3) and provides sufficient services to support their work (mode, median values equal 3), whereas they are willing to recommend it to their community (mode, median equal 3). They are more reluctant to use the Dicode integrated services in the near future (mode, median equal 2) or change their current work practices (mode, median values equal 3 and 2, respectively). This can be partly attributed to the reasons mentioned above.

8.5 Conclusion

The CGRA case concerns multidisciplinary biomedical research communities, ranging from biologists to bioinformaticians, which need to collaborate in order to assimilate clinico-genomic research information and scientific findings, and explore diverse associated issues. In many cases, such collaboration needs to take into account very large datasets, emphasizing the need of well-established practices which assist scientists to understand how to manage, navigate and curate large-scale data.

The Dicode approach enables a meaningful aggregation and analysis of large-scale data in complex settings, such as that of biomedical research. The proposed solution allows for new working practices that turn the problem of information overload and cognitive complexity into the benefit of knowledge discovery. This is achieved through a properly structured information network that can be used as the basis for more informed decisions. Simply put, the Dicode approach is able to turn information growth into knowledge growth; it improves the quality of collaboration's output, while enabling users to be more productive and focus on creative activities.

References

1. Tsiliki, G., Kossida, S.: Fusion methodologies for biomedical data. J. Proteom. **74**, 2774–2785 (2011)
2. Karacapilidis, N., Tzagarakis, M., Christodoulou, S., Tsiliki, G.: Facilitating and augmenting collaboration in the biomedical domain. Int. J. Syst. Biol. Biomed. Technol. **1**(1), 52–65 (2012)
3. Sullivan, D.E., Gabbard, J.J., Shukla, M., Sobral, B.: Data integration for dynamic and sustainable systems biology resources: challenges and lessons learned. Chem. Biodivers. **7**(5), 1124–1141 (2010)
4. Koschmieder, A., Zimmermann, K., Tribl, S., Stoltmann, T., Leser, U.: Tools for managing and analyzing microarray data. Brief Bioinform. **13**, 46–60 (2011)
5. Lukk, M., Kapushesky, M., Nikkilä, J., Parkison, H., Goncalves, A., Huber, W., et al.: A global map of human gene expression. Nat. Biotechnol. **28**, 322–324 (2010)
6. Joyce, A.R., Palsson, B.Ø.: The model organism as a system: integrating 'omics' data sets. Nat. Rev. Mol. Cell Biol. **7**, 198–210 (2006)
7. Pennisi, E.: Will computers crash genomics? Science **331**, 666–668 (2011)
8. Huttenhower, C., Schroeder, M., Chikina, M.D., Troyanskaya, O.G.: The Sleipnir library for computational functional genomics. Bioinformatics **24**(13), 1559–1561 (2008)
9. Baker, M.: Quantitative data: learning to share. Nat. Meth. **9**, 39–41 (2012)
10. Guberman, J.M., Ai, J., Arnaiz, O., Baran, J., Blake, A., Baldock, R. et al.: BioMart central portal: an open database network for the biological community. Database bar041 (2011)
11. Perez-Llamas, C., Gundem, G., Lopez-Bigas, N.: Intregative Cancer Genomics (IntoGen) in Biomart. Database bar039 (2011)
12. Lee, S.E.: Facilitating collaborative biomedical research. In: Proceedings of GROUP '07. Sanibel Island, Chicago, USA (2007)
13. Finholt, T.A.: Collaboratories as a new form of scientific organization. Econ. Innov. New Technol. **12**(1), 5–25 (2003)
14. Spencer, D., Zimmerman, A., Abramson, D.: Special theme: project management in E-Science: challenges and opportunities. Comput. Sup. Coop Work **20**, 155–163 (2011)
15. Atzmueller, M., Puppe, F., Buscher, H.P.: Exploiting background knowledge for knowledge-intensive subgroup discovery. In: Proceedings of IJCAI'05, pp. 647–652 (2005)
16. Huber-Keener, K.J., Liu, X., Wang, Z. et al.: Differential gene expression in Tamoxifen-Resistant breast cancer cells revealed by a new analytical model of RNA-Seq data. PLoS ONE. 7, 7, e41333. DOI= doi:10.1371/journal.pone.0041333 (2012)
17. Kitchenham, B., Pfleeger, S.L.: Principles of survey research part 5: populations and samples. Softw. Eng. Notes **27**(5), 17–20 (2002)
18. Cugini, J., Damianos, L., Hirschman, L., Kozierok, R., Kurtz, J., Laskowski, S., Scholtz, J.: Methodology for Evaluation of Collaboration Systems, The Evaluation Working Group of The DARPA Intelligent Collaboration and Visualization Program Revision 3.0 (1997)
19. Sun, Y., Greenberg, S.: Places for lightweight group meetings: the design of come together. In Proceedings of GROUP'10, Sanibel Island, Florida, USA, pp. 235–244 (2010)
20. Nielsen, J.: Designing Web Usability: The Practice of Simplicity. New Riders Publishing, Indianapolis (1999)
21. Norman, D.A.: The Design of Everyday Things. The MIT Press, London (1998)

Chapter 9
Opinion Mining from Unstructured Web 2.0 Data: A Dicode Use Case

Ralf Löffler

Abstract Web 2.0 is challenging existing marketing and communication paradigms. Social Web has given the consumers a voice and Social Media has huge impact on brands and products today. To better understand the customers, Social Media Monitoring has become the most important way to listen to their voices. Tools available today are critically questioned from marketing regarding comprehensiveness and truth as well as regarding its representativeness. The lack of internal marketing resources involves third parties into the Social Media Monitoring process. Dicode's goal is to support a collaborative work environment and offer technical solutions that improve the overall quality in the social media processes for all parties involved. This chapter reports on the use of Dicode Workbench and Dicode's services in the above context.

Keywords Social media · Marketing · Social media monitoring · Big data analysis

9.1 Web 2.0: Social Media Monitoring in Marketing

9.1.1 A New Communication Paradigm and the New Power of Customers

Web 2.0 is heavily challenging existing marketing and communication paradigms and is an incredibly dynamic environment. For a long time, brands communicated in a top-down mode: firms were the senders, while consumers were the receiver or

Formerly, Ralf Löffler was the Managing Director of Publicis Brand Consultancy. Currently, he is a co-founder and Partner of the ISK Institut for Strategy and Communication.

R. Löffler (✉)
ISK Institute for Strategy and Communication, Frankfurt, Germany
e-mail: ralf.loeffler@isk-institut.de

N. Karacapilidis (ed.), *Mastering Data-Intensive Collaboration and Decision Making*, 181
Studies in Big Data 5, DOI: 10.1007/978-3-319-02612-1_9,
© Springer International Publishing Switzerland 2014

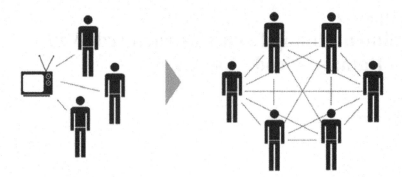

Fig. 9.1 Radical change in communications; from "sender-receiver principle" to "cross-linked communications"

addressees (Fig. 9.1, left part). These brands controlled the conversation in terms of topics, critics and media penetration. Today, such a model of communication is obsolete; instead, a cross-linked communication model has been adopted (Fig. 9.1, right part). Accordingly, brand managers from agencies and companies must reassess and realign their tools in order to cope with the rise of the Web 2.0 era.

Web 2.0 (Social Media) has given customers a voice; they communicate at all times and everywhere, build networks, spread their opinion, generate their own content and continuously demand news and interaction. Consumers have a huge impact on brands and products now. The fusion of World Wide Web and mobile phones changed people's communication behaviour and habits. New opportunities made them even more permissive, demanding and self-confident. They actively influence the perception of the brand: By now, 80 % of online content is user generated, which outnumbers the branded communications by companies and agencies [1].

9.1.2 The Social Web

The Social Web has become huge, diverse and is permanently changing. The input for the intended Social Media Monitoring Process comes from it. There are basically two categories that are used as input: professionally published online content and user generated content (Social Web). The Social Web, in particular, is very unstructured, loud and dynamic. The amount of available data is growing extremely fast as it is based on the high growth rates of the web. Social Media comprises very different types of web services such as Blogs, Social Networks, Micro Blogs, Forums, Websites, Online News, Video Sharing Platforms, Picture and Music Exchange Platforms, Rating Sites, Wikis, etc. (Fig. 9.2).

Fig. 9.2 The conversation prism (*Source* Brian Solis, http://www.briansolis.com/2010/10/introducing-the-conversation-prism-version-3-0/)

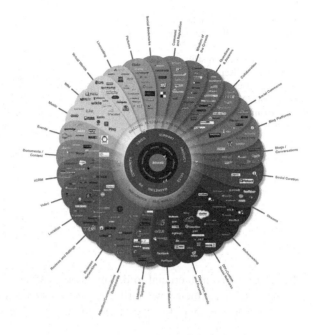

9.1.3 Big Data: Its Meaning for Marketing and Marketing Consulting

Brand managers of companies not only have to take the web into consideration (if they don't consider it, they jeopardize the success of their brands and companies); the consumers' voice is much more reliable than that of the company or traditional communication channels like TV/print/outdoor (Fig. 9.3). In other words, the consumers have taken over the power.

This forces companies to invest in additional resources with the aim of understanding the "whom", "what" and "where" in the Social Web. With Social Media Monitoring (SMM), Marketing and Strategic Planning have a complex and versatile tool for listening to the Internet and the consumer's voice. All comments related to brands and companies, reviews of products or conversations between consumers about brands can be located and filtered by utilizing appropriate software tools. For this purpose, a variety of services with different priorities and qualities have been developed in recent years. Some are freely available (e.g., Google Blog Search or Technorati), while a fee is required for others (e.g., Radian6, Sysomos and SM2). Thus, SMM tools seem to be the right solution to rebuild a company's capacity to listen to and interact with consumers.

Today, companies do have an enormous need for data to improve their business steadily. At the same time, they usually do not have the capacities and/or the knowledge to keep up with current innovations and develop their own solutions. That's why the data business, especially for marketing, is a consultancy business.

Fig. 9.3 To what extend do you trust the following forms of advertising? (*Source* Nielsen "Global Trust in Advertising" Report. http://www.fi.nielsen.com/site/documents/NielsenTrustinAdvertisingGlobalReport April2012.pdf)

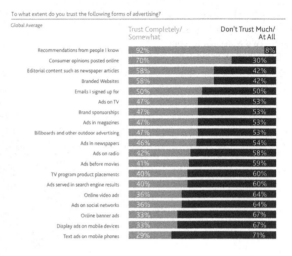

For consultants, it is crucial to deeply understand new technologies and ways of thinking to keep their own business up to speed and serve their clients with "customer-made data" to support them in achieving better business results.

9.1.4 The Internet's Role for Big Data Analysis in Marketing and Research

Size. The biggest information source for these "data junkies" is the Internet. Web 2.0 is by far the biggest and at the same time the most aggressively growing data source on the planet. Every two years, the amount of data on the Internet almost doubles. It is a hyper-complex, fast-growing and fast-changing space. It is not possible to overview it without any software tools. The daily uploads on YouTube alone amount to about 144,000 hours. And YouTube is only one website, one forum out of millions. Internet users not only use data, they produce data in the same way. "Data users" have become "data producers". This produces a "data-tsunami", which is a very interesting source for those who want to take a look at the data and obtain relevant information from it.

Influence. To find out relevant aspects about a theme, the Internet is absolutely essential. The first thing people do to inform themselves about something is to "ask" the Internet; Internet is *the* opinion-maker. Influencers are the most relevant group in terms of evaluation and more important than ever before. That is why an observation of them is crucial when it comes to research. To a certain degree, SMM tools are able to find the influencers on a quantitative basis. Finding influencers is one of the most difficult parts of the Social Media analysis. The other—not less important—aspect is the opinion leader change due to the use of the Internet. This year, the Grimme Online Award in Germany was awarded for the very first time to a Twitter-hashtag; one word, spread out by one person was caught up by thousands of people on the Internet and even influenced the talk-shows on TV (http://www.grimme-institut.de/html/index.php?id=1667#c10914). Not knowing what is happening on the Internet means not knowing what happens at all.

Research Conditions. The Internet is the place where you can obtain genuine opinions from real people without the problem of distortion effects in traditional market research. The Internet is a transparent room of opinion clusters that directly concern the perception of things: brands, decisions, news, politics, etc. Opinions can travel very quickly through the Social Web. For Marketing, you need this data in real-time. It is extremely difficult to find all relevant opinions about your category and brand. Terms like "friend" and "influencer" are no longer adequate to describe the scope of social activity and interaction in the current Social Web. A deep understanding of consumer needs and motivations is the key to unlocking a real understanding of Social Media and its users.

That is why SMM has become a necessity. To make big data useful for marketing, there is actually no other valid possibility. Besides allowing for quantitative data pooling, the tools can capture data with a very specific theme in a specific time frame. This forces companies to invest in additional resources to understand the who, the what and the where in the Social Web. A big worldwide survey from the Meltwater Group (involving 450 Companies) shows that more than 84 % of the participating companies would like to invest in monitoring such data, but less than 20 % actually do (http://melwaterproducts.com/reports/melwater_future_of_content_report.pdf). Two reasons appear to be the strongest for that: (i) the quality of the existing solutions is not sufficiently reliable to build a solid basis for marketing and business strategies, and (ii) integration of Social Media Monitoring into a company's existing structure and marketing procedures is often difficult.

9.2 Major Challenges for Marketing

Nowadays, every company has to find out how their customers talk about their company and their brands and about its competition. Marketing departments use Social Media Monitoring tools to observe what the Internet "says", because their target groups talk on the Internet and influence the perception of things. Hence, they are dependent on relevant data to better understand their target audience and act/react very fast. These tools can help to obtain the desired data but it is always the question whether the data tells the truth and is comprehensive or meaningless because the detected data represents only a "snippet" of what is potentially available. Many companies and their managers are not happy with the results and some of them have stopped working with the tools because of dissatisfying and non-plausible results.

Secondly, because of lacking internal resources, they have to involve external highly qualified experts/companies to be able to obtain all the input needed for the decision-making process. Hence, consulting hedges or backs up their decision. Therefore, the quality of data and the "intelligence" of tools are the keys of

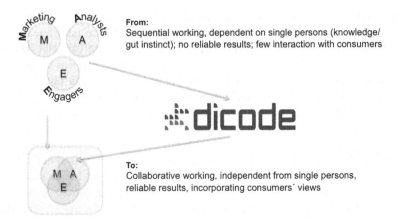

Fig. 9.4 From a sequential working environment to a collaborative infrastructure

success for the consulting business. Working with third parties addresses this second important challenge: the need for a working environment that successfully involves different parties into the Social Media Monitoring (Fig. 9.4).

9.3 The Dicode Solution

The Dicode project aimed to facilitate and augment collaboration and decision-making in data-intensive and cognitively-complex settings. The *Use Case* reported in this chapter focuses on supporting opinion mining from unstructured Web 2.0 data involving different parties in the Social Media Monitoring Process. The ability of Dicode's integrated services to analyze complex texts has impact on the marketing decision-making processes, creating opportunities to deliver by far quicker and better analytical results, and at the same time creating competitive advantages for brands/products.

Within companies and the marketing departments, there is a need for a (r)evolution. The sheer amount of mails, mailing lists, conferences calls, reports, video conferences, etc. is seen as an inadequate way of working together in digital times. Often, it provokes similar working operating cycles and process ruptures, which lead to ineffectiveness and inefficiencies. A very similar effect is produced by working with third parties: no common place to work, no common space to share information, no platform offering a complete and comprehensive overview for all parties involved in a project regarding the project status and the collaborative work. As a result, collaborative work is very often a good idea rather than real action.

The Dicode Workbench (see Chap. 7) facilitates significantly collaborative work and decision-making processes. It is a web application that provides a common graphical interface to access and use heterogeneous services. It makes the sharing and exchange of information (files, reports, etc.) in data-intensive and cognitively-complex settings more comfortable.

9.4 Working with the Dicode Platform

This section presents two scenarios concerning the use of the Dicode Workbench and Dicode Integrated Services for the analysis of the voluminous amount of unstructured information existing on the Web (especially, in the highly dynamic Social Media space). We show how different parties from different locations can work together effectively using the above mentioned Dicode solutions.

Three people are involved in the Use Case reported, namely: *Frank* (Brand Manager in the automotive industry, Stuttgart), *Alice* (Social Media Analyst, Hamburg), and *Nathalie* (Social Media Engager, Berlin). The Dicode services incorporated in the scenarios below are: *Collaboration service* (see Chap. 6), *Top Entity service* (see Chap. 5), *Prominence Graph service* (an earlier version of the *Entity Prominence service* described in Chap. 5), *Phrase Extraction service* (see Chap. 5), and *Topic Detection service* (see Chap. 5).

9.4.1 First Scenario: Opinion Mining

The first scenario provides insights on the following questions:

• What do people do to analyse the Web 2.0? How do they do it and why?
• How do people work together from different places?

Once a week, Frank uses the Top Entity service to check some relevant buzz (Fig. 9.5). He discovers heavy discussions about his brand within a specific time frame and wants to find out more about them (what is the reason for this buzz?). Frank opens a new Collaborative Workspace (offered through the Dicode Workbench from the Dicode's Collaboration service) and contacts Alice. She is asked to investigate where the buzz comes from. Alice reads Frank's message in the Collaborative Workspace. She sends Frank a short reply and starts with the project.

First, Alice uses the Prominence Graph service to identify peaks within a specific keyword and timeframe. There are two interesting peaks for the keyword "Mercedes-Benz" (Fig. 9.6). By clicking on the peak, Alice is automatically forwarded to the Google results for this specific day. By checking the links, she notices a lot of entries related to the car brand to the MB Fashion week. Then, she double-checks with the peaks for the main competitors: "Audi" and "BMW" (just to exclude the case that the peak is generated from the whole set). She observes that it is a Mercedes Benz peak. Finally, she writes a message to Frank that she has

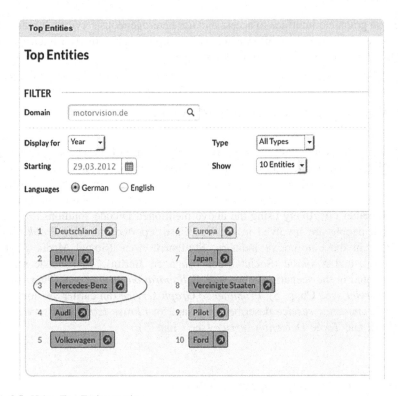

Fig. 9.5 Using Top Entity service

found the reason for the buzz and recommends extending the collaboration with the Fashion Week. Then, she connects her results with Frank's initial message.

Frank checks the message on the Collaborative Workspace and is very impressed with the results. He really likes the idea of boosting the collaboration with the Fashion Week and writes a comment on the Workspace to Alice and Nathalie. Nathalie is the Social Media Engager and she should develop the Social Media strategy for the collaboration. Again, Frank connects his comment with Alice's comment on the Workspace (Fig. 9.7). Nathalie sees the conversation, writes a short note to Frank and starts working on the project.

9.4.2 Second Scenario: Deep Analysis of Complex Texts

This scenario provides insights about how the Dicode services can help to deeply analyze complex texts and collaboratively make decisions.

Frank wants to know how the newly launched "A-Class" is perceived and discussed on the Web to adjust communications. Alice collects car reviews from the "autobild" (online car magazine) as a starting point for her analysis. She

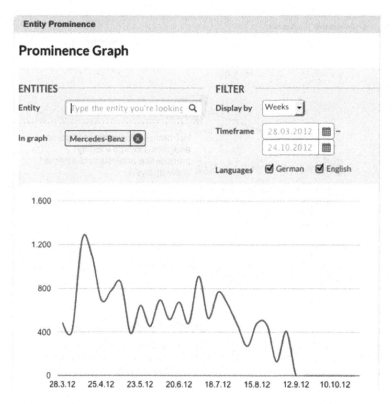

Fig. 9.6 Using Prominence Graph service

arranges the texts in directories, which are named after the cars that are reviewed. Then, she zips the directories into a single file. Alice uses the Topic Detection service (Fig. 9.8) to quickly understand what the conversations are about and which sub topics drive the conversations.

Alice executes the service from the Dicode Workbench. She chooses to do a fine-grained analysis (with many topics). She wants to get the graphical results display in a PDF file. She specifies the text language: German. Alice also wants to see the car types in the resulting graph: therefore she tags the "display text categories" button and indicates that she only wants to see the 10 most prominent car types. Alice does not want to see all topics of the fine-grained analysis. Finally, she triggers the execution of the service.

The analysis uncovers the A3 as a close competitor (Fig. 9.9). In addition, handling, performance and consumption has been identified as the most competitive topics (Fig. 9.10).

To better understand the conversations and the drivers in the category. Alice uploads two review texts. She wants to detect only positive emotions in the texts. She indicates German as the text language. She chooses a predefined emotional analyzer for automotive texts (Fig. 9.11). In addition, Alice wants to see the two

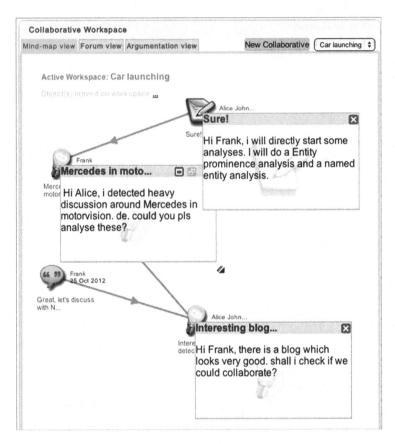

Fig. 9.7 Frank and Alice share their knowledge with Natalie

texts with the positive phrases highlighted (Fig. 9.12). She can also see a tag cloud, which is generated from the positive phrases in the two texts (Fig. 9.13).

The results are discussed with the whole team. If the results are not satisfactory in terms of relevance or if they are not plausible, the model can be improved by providing further input. After seeing the results, Frank made a decision and contacts all team members through the Collaborative Workplace. He decides to push the new "A-Class" at the next Mercedes-Benz Fashion Week.

9.5 Feedback from Industry

To improve the usability and practical benefits of the Dicode solution, two evaluation rounds were conducted. During the second evaluation round, high-level marketing professionals (real users) were questioned in March and April 2013 to evaluate the enhanced version of the Dicode Workbench and Dicode services that

Topic Detection

	upload ZIP archive ☑ sub directories for categories	)eliverables\D6.4.2\Fahrbericht.zip Durchsuchen..
Text Collection Source:	○ Query text harvester	time interval: Start date: 01 ▾ 04 ▾ 2011 ▾ End date: 02 ▾ 04 ▾ 2011 ▾ collection type: ○ Twitter ⦿ Blogs
☐ display terms of interest	Terms of interest file:	Durchsuchen..
Topic Model Parameters:	○ build a coarse-grained topic model ⦿ build a fine-grained topic model	
Graph Format:	⦿ provide PDF file ○ provide Gephi file	
Specify language:	○ English ⦿ German	
Other options:	☑ Display text categories - number of categories to display: 10 ☐ Display terms of interest - number of terms to display: 10 ☑ Select top topics - number of topics to display: 20	
Execution:	Go!	
Cancel:	[X] Close Window	

Fig. 9.8 Using the Topic Detection service

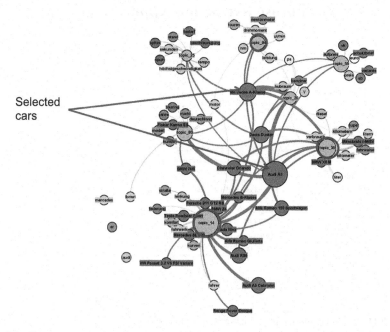

Fig. 9.9 The Topic Detection service shows relations between brands and topics based on big data text input. Audi A3 has been identified as close competitor

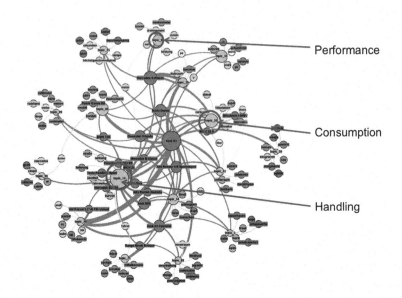

Fig. 9.10 Output of the Topic Detection service which has identified three very competitive topics

Phrase Extraction Application

Text Collection Source:	● upload ZIP archive	...ng\Deliverables\D6.4.2\Fahrberic Durchsuchen...
	○ Query text harvester	time interval: Start date: 01 ▾ 04 ▾ 2011 ▾ End date: 02 ▾ 04 ▾ 2011 ▾ collection type: ○ Twitter ● Blogs
Model Type:	● Automotive texts ○ Laptop texts	● Detect positive phrases ○ Detect negative phrases
	○ upload model	Durchsuchen...
Specify language:	○ English ● German	
Other options:	☑ Display new phrases after completion ☑ Show tag cloud after completion	
Execution:	Go!	
Cancel:	[X] Close Window	

Fig. 9.11 Working with the phrase extraction application

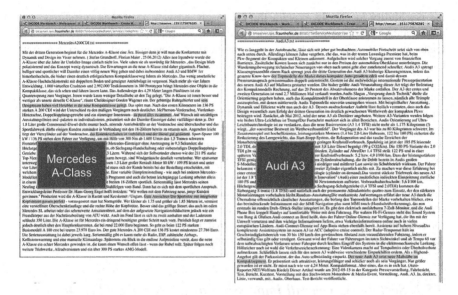

Fig. 9.12 Car reviews with positive *phrases highlighted*

Fig. 9.13 The tag cloud shows the positive phrases of the car reviews from "A-Class" and "Audi A3"

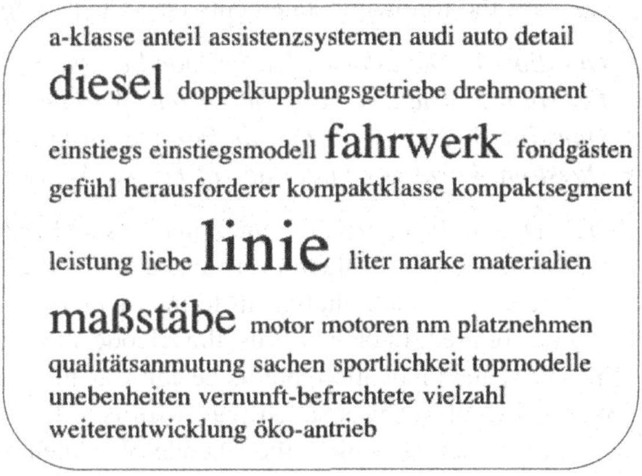

a-klasse anteil assistenzsystemen audi auto detail diesel doppelkupplungsgetriebe drehmoment einstiegs einstiegsmodell fahrwerk fondgästen gefühl herausforderer kompaktklasse kompaktsegment leistung liebe linie liter marke materialien maßstäbe motor motoren nm platznehmen qualitätsanmutung sachen sportlichkeit topmodelle unebenheiten vernunft-befrachtete vielzahl weiterentwicklung öko-antrieb

are related to the particular Use Case. The research performed had two main objectives: to identify importance and challenges of Social Media Monitoring in a corporate context, and to ensure the relevance and usability of the Dicode Workbench and Integrated Services. Parallel to that, telephone interviews with selected marketing professionals were carried out.

For the interviewed experts, the "big picture" and the easy usability is by far more important than details. Their expectation is more to be faced with a nearly final or with an "easy to understand" version of the proposed system. The research was carried out through online questionnaires, accompanied by informative video-casts, as well as through individual telephone interviews, in order to have the possibility to clearly explain the approach, answer upcoming questions, and obtain a deep understanding of their needs and judgments. All evaluators were in leadership positions in industry, communication agencies and universities with huge experience in the field of digital communication.

In addition to rating the statements/questions of the online questionnaire, evaluators had also the opportunity to comment on them. There was almost no criticism on the actual concept of the Dicode Workbench and its Integrated Services. Generally speaking, the overall results were very satisfying. In what follows, we present the quantitative and qualitative feedback derived by the online questionnaires, divided into three sections (Sects. 9.5.1, 9.5.2 and 9.5.3). Answers to the quantitative questions of the questionnaires were given for ordinal data in a 1–5 scale (questions concerning the quality, acceptability and accessibility of the services provided), where 1 stands for 'I strongly disagree' and 5 for 'I strongly agree', and for continuous numerical data (scale data) in a 0–10 scale (questions concerning the services' usability), where 0 stands for 'none' and 10 for 'excellent' [2, 3].

9.5.1 Section A: Overall impression of Dicode Workbench

In this section, evaluators were asked to answer how strongly they agree or disagree with the following statements (see Chap. 8 for related responses):

- *Question 1*: The information provided on the video is clear.
- *Question 2*: The design of the Dicode Workbench is very pleasant.
- *Question 3*: The use of Dicode Workbench is easy.
- *Question 4*: The user interface of Dicode Workbench is intuitive.

The Dicode Workbench's functionalities and usage seemed to be easily adoptable. However, most evaluators criticized the "old-fashioned" style of the interface. It was seen as a contradiction in terms of innovativeness compared to the functionalities offered. In other words, the Dicode Workbench was appreciated in terms of newness and innovation, but its design was judged as unpleasant/old fashioned / not user friendly. It was stated that there is still room for improvement as far as switching windows within the Dicode Workbench is concerned. As shown in Fig. 9.14, the mode and median values of Section A's statements vary from 2 to 4. (The comments concerning functionality and design had been addressed within the final version of the Dicode Worbench.)

9.5.2 Section B: Support of Collaboration, Decision Making and Data-Mining

In this section, evaluators were asked to rate their agreement with respect to the following statements:

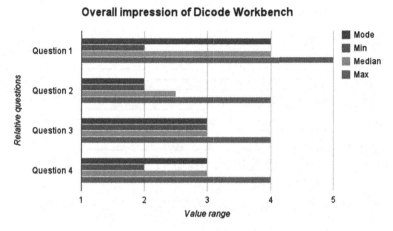

Fig. 9.14 The overall impression of Dicode Workbench: mode, minimum, median, and maximum values are presented for questions *1–4*. The range of original values is *1–5*, where *1* denotes strong disagreement and *5* denotes strong agreement

- *Question 5*: The "Topic Graph" service is very helpful to easily indentify competitive topics.
- *Question 6*: The "Top Entity Service" allows me to easily discover the discussions about my topic (in a certain type, for a certain domain, at a certain point in time).
- *Question 7*: The "Named Entity Service" discovers precisely the right object of my investigation (e.g. liquid spice named "Maggi" and not the first name "Maggi").
- *Question 8*: The "Prominence Graph" quickly finds the entity occurrences over time and in comparison to competition.
- *Question 9*: With the "Prominence Graph and Google", I easily understand the drivers of the conversation.
- *Question 10*: The "Topic Graph" is very helpful to easily identify competitive topics.
- *Question 11*: The "Phrase Extraction Application" is an automatic learning service which strongly supports my update skills.
- *Question 12*: Sentiment Analysis has become an easy and time saving task by using the "Phrase Extraction Application".
- *Question 13*: The automatic highlighting of positive/negative phrases within a certain context in the "Phrase Extraction Application" is very valuable.
- *Question 14*: The easiness to analyse complex text will strongly support the speed of operation and collaborative working.
- *Question 15*: Dicode services can help me to deal with cognitive-complex issues.
- *Question 16*: The Dicode Workbench can facilitate collaboration.
- *Question 17*: The Dicode Workbench can enhance decision making.

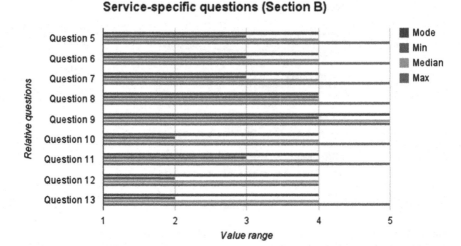

Fig. 9.15 Support of collaboration, decision-making and data-mining: mode, minimum, median, and maximum values are presented for services' specific questions *5–13*. The range of original values is *1–5*, where 1 denotes strong disagreement and *5* denotes strong agreement

- **Question 18**: *The services of Dicode Workbench are very well integrated.*
- **Question 19**: *The Dicode Workbench can help me be more productive and concentrate on creative activities.*

As shown in Fig. 9.15, the individual Dicode Services (Questions 5–13) are rated with a median of 4 and even 5 in one case ("Prominence Graph and Google"). This is a very satisfying result that indicates the actual relevance and the proper development of the services. When considering a more general view of the integrated services (Questions 14–19), experts were a little more skeptical with their rating. Especially in terms of dealing with cognitive-complex issues, integration, creative activities and decision-making. Nevertheless, the mode and median values range from 3 to 5 (Fig. 9.16), which indicates their overall acceptability.

9.5.3 Section C: Potential Benefit to My Work (Willingness to Use the Dicode Workbench in Their Working Environment)

In the last section of the questionnaire, evaluators were asked to express their willingness to use or recommend Dicode Workbench, as well as to change their current work practices. The statements of Section C were:

- **Question 20**: *The "Topic Detection" and the "Topic Graph" services will be a strong support for my analytical work.*
- **Question 21**: *I can see the potential benefit of using Dicode Workbench in my work.*

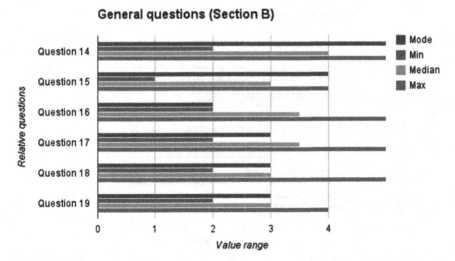

General questions (Section B)

Fig. 9.16 Support of collaboration, decision-making and data-mining: mode, minimum, median, and maximum values are presented for services' specific questions *14–19*. The range of original values is *1–5*, where *1* denotes strong disagreement and *5* denotes strong agreement

- **Question 22**: *Dicode provides sufficient services to support my work.*
- **Question 23**: *I would consider using the Dicode Workbench in the near future.*
- **Question 24**: *The use of Dicode will have positive impact on my current work practices.*
- **Question 25**: *The use of Dicode will change my current work practices.*
- **Question 26**: *I will recommend the Dicode Workbench to my peers/community.*

A positive impact of the Dicode Workbench and Dicode Services on their current work practices in the near future was given by the majority of the interviewees. The readiness to recommend Dicode is on a high level. In Fig. 9.17, we can observe that the mode and median values range in (2, 4), although most values are within the (3, 4) range. Even if the experts made very positive comments on the Dicode Workbench and the Dicode Services, the reaction on their use and recommendation was slightly more reserved. This can be explained by the fact that many experts stated in the personal interviews that they really need to use the tools to form a final opinion. Nevertheless, Dicode services were evaluated as highly relevant for marketing today in order to derive meaningful information from the Web; they address important marketing challenges and strongly support Social Media Monitoring.

9.6 Opportunities in Marketing

The addressed key target groups of Dicode's Workbench and Integrated Services are communication agencies (e.g. Advertising, PR, Social Media, Media), Social Media Monitoring tool providers and companies/multinational corporations.

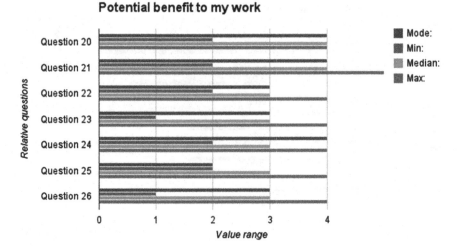

Fig. 9.17 Potential benefit to my work: mode, minimum, median, and maximum values are presented for services' specific questions *20–26*. The range of original values is *1–5*, where *1* denotes strong disagreement and *5* denotes strong agreement

9.6.1 Communication Agencies

Today, most of these agencies develop, recommend or plan online communication without really knowing what the target groups conversations are about. While they talk about 360° communication, they still have no or far too little information about the target groups' communicational behaviour, their attitudes and the content of the conversations. Furthermore, the online media strategy recommended by various affiliations represents only the leading online marketers in Germany. This is truly only a small proportion of the online opportunities. Clearly spoken, their recommendation is based on what they know today and not on what is really happening today. All agencies have to develop content that is inviting, interesting and entertaining for their audience. By having only limited knowledge about their interests, how can they develop exciting content for them?

All agencies talk "strategy". Strategy is the description of the path from A to B. How do they develop their Internet strategy without knowing how they are seen in the eyes of their audience and then develop the route to B? The Dicode services may offer huge opportunities to better understand the brand's audiences in Web 2.0; they may also support communication agencies to better understand the target groups and generate insights as a valid basis for future communication strategies.

9.6.2 Companies/Multinational Corporations

Similar to communication agencies, companies have a fundamental interest in knowing what their audience thinks about them and their brands. They need a "seismographic" tool that is able to recognize slight changes regarding their brand in the buzz. They also want to get to know how the Internet talks about the competitors. Until now, they aren't getting what they need to know. This is due to the fact that only a few companies consider SMM as an opportunity to listen and discover insights.

Most companies still use Social Media as a PR channel or something they can use like a New Media channel. Those who are already aware of the power of Web 2.0 are very dissatisfied with the current tool providers, because they really don't know whether they are receiving comprehensive data. Also because the tools are far too slow, not accurate and still not state-of-the-art. They see that the tools available today are far behind the development of the Internet.

In contrast, Dicode represents the clear opposite: comprehensive, pioneering innovations and inventions, fast and reliable. Dicode's Collaborative Workspace and Dicode Services allow companies to really speed up to the needs of Web 2.0. Dicode is able to not only offer tools but also to consult companies to understand the digital opportunities and do not see Web 2.0 as a New Media channel only. The Dicode Collaborative Workspace is a perfect platform for companies to work with third parties efficiently and effectively.

9.6.3 Social Media Monitoring Providers

They provide their clients with data. They are very quantitatively driven and are just beginning to understanding clients and consumers. What their clients are asking for is far beyond that; in addition to data, they want to get analytical tools, self-learning systems and real insight into the lives of consumers. Making them better is a big and easy to realize business model for exploiting the overall Dicode approach. The Named Entity service, the Topic Detection service and the Phrase Extraction service are tools that ideally fulfil the needs of the respective client.

9.7 Conclusion

Dicode's Workbench and integrated services are able to support collaborative work and decision making in an extremely fast changing environment of Web 2.0. Analysing complex texts, creating competitive advantages, delivering quicker and by far better analytical results are the obvious advantages of Dicode. Despite all the technical benefits of Big Data analysis, one should always consider the linkage to human beings, their needs and desires. The best given data is only as good as the expert who analyses the data. Thus, all developments need to be human centred at the first stage. This is fully in line with the overall Dicode approach, which builds on the synergy of human and machine reasoning.

References

1. Tenenbaum, I.: Brands that dominate with user-generated content. iMedia Connection Report. http://www.imediaconnection.com/content/34502.asp#multiview (2013)
2. Nielsen, J.: Designing Web Usability: The Practice of Simplicity. New Riders Publishing, Indianapolis (1999)
3. Norman, D.A.: The Design of Everyday Things. The MIT Press (1998)

Chapter 10
Data Mining in Data-Intensive and Cognitively-Complex Settings: Lessons Learned from the Dicode Project

Natalja Friesen, Jörg Kindermann, Doris Maassen
and Stefan Rüping

Abstract This chapter reports on practical lessons learned while developing the Dicode's data mining services and using them in data-intensive and cognitively-complex settings. Various sources were taken into consideration to establish these lessons, including user feedbacks obtained from evaluation studies, discussion in teams, as well as observation of services' usage. The lessons are presented in a way that could aid people who engage in various phases of developing similar kind of systems.

Keywords Data mining framework · Data mining services · Text mining services · Big data · Hadoop · Storm · Semantic technologies

10.1 Introduction

A major concern of dealing with Big Data is the dichotomy between latency and scalability. Analysis of huge amounts of data takes time, which implies high latency, while obtaining analyses in short time is typically feasible only for small data with bad scalability. The Lambda Architecture [1] has been proposed as a

N. Friesen (✉) · J. Kindermann · S. Rüping
Fraunhofer IAIS, Schloss Birlinghoven, 53754 Sankt Augustin, Germany
e-mail: friesen.natalja@googlemail.com

J. Kindermann
e-mail: joerg.kindermann@iais.fraunhofer.de

S. Rüping
e-mail: stefan.rueping@iais.fraunhofer.de

D. Maassen
Neofonie GmbH, 10115 Berlin, Germany
e-mail: doris@neofonie.de

N. Karacapilidis (ed.), *Mastering Data-Intensive Collaboration and Decision Making*, 201
Studies in Big Data 5, DOI: 10.1007/978-3-319-02612-1_10,

compromise combining scalable and robust analysis of Big Data using Hadoop with real-time analysis, e.g. Storm. These developments have been so recent that they were beyond the scope of Dicode, but have and will have some impact on exploitations of the Dicode results. Dicode has focussed on improving scalability—and to some extend latency—using the Hadoop-based approaches to data-centred parallelisation. Lessons learned with regard to this technology are listed in Sect. 10.2.1. Scalability and latency do not only depend on the technology used but also on the services offered. Lessons learned with regard to services are subsumed in Sect. 10.2.2. Finally, the effectiveness of analyses not only depends on technology and services, but also on appropriate and efficient usage of these. Section 10.2.3 discusses lessons learned from this perspective.

10.2 Lessons Learned

The lessons discussed below concern experiences, concrete recommendations and best practices related to the process of developing the infrastructures and services for mining data.

10.2.1 Technology

10.2.1.1 Lesson 1: Innovative Big Data Solutions Can Greatly Simplify Data Processing Tasks

In Dicode, we used Wikipedia as a central resource for Named Entity Disambiguation. From Wikipedia's link structure, we derived the probabilities of signifiers for certain entities. Amongst other statistics, we count links from a certain surface form to a Wikipedia article (e.g. how often "George Bush" either links to the father or to the son).

With 4.2 Million articles, the German Wikipedia seems quite small from a Big Data perspective. For our analysis, each article is analysed several times, because global information like N-gram statistics is needed and recursive link processing is required. Before Hadoop was introduced, the Wikipedia analysis component was implemented as a sequential UIMA pipeline (http://uima.apache.org/). In total, the analysis took several days on a standard machine. In Dicode, the processing time was reduced to several hours. Now a *cron job* (http://en.wikipedia.org/wiki/Cron) regularly checks if a new Wikipedia dump has been published and generates the statistics automatically. This lesson shows us that our idea of "size" changed tremendously during the last three years. Before Dicode, processing a collection of several million documents took a couple of days. Today, this is reduced to a couple of hours or even minutes. In addition, configuration-intensive frameworks like

UIMA today are often replaced by more light-weight approaches. Developing Big Data solutions in most cases does not require writing complex MapReduce jobs. With Apache Pig, an easy to use scripting language is available, which allows for rapid development. Dicode's new Wikipedia statistics component is pretty concise: it contains about 100 lines of code (Wikipedia statistics are computed by the pig script appearing at: https://github.com/Dicode-project/pignlproc/blob/master/examples/nerd-stats/nerd-stats.pig).

10.2.1.2 Lesson 2: Conducting a Big Data Project Without a Shared Cluster Infrastructure Slows Down Development

From the beginning of Dicode, it was clear that it was not feasible to rely on a shared hardware infrastructure designed for Big Data: none of the partners already had an appropriate cluster infrastructure at hand and there was no explicit hardware budget included in the project calculation. Each of the partners therefore set up a separate development infrastructure.

An important question was how to store and access commonly used data sets. Batch processing systems like Hadoop achieve high scalability and throughput by moving the computation to the data: algorithms are executed on a subset of data that is stored locally at the individual nodes and the results of those analyses are later collected and assembled. We first experimented with a combined approach: for development purposes, a small subset of documents was downloaded by the respective partner and subsequently used on their local infrastructure, e.g. for training of machine learning models. Later, the developed component was wrapped into a user defined function (UDF) and executed on the cluster. We used this rather cumbersome approach in the development of an early version of Dicode's *Named Entity service* (see Chap. 5), which was developed by Fraunhofer IAIS (FHG) based on Conditional Random Fields. FHG had no direct access to the cluster of Neofonie GmbH (NEO), because a shell access to the development environment does not comply with the company's security standards. Additionally, FHG uses a different technology stack on their Hadoop infrastructure. The integration of the algorithms into a UDF was therefore performed by NEO, which leads to longer development cycles. To avoid this in the future, we would opt for a shared hardware infrastructure, for which an extra budget and investment into a secure set-up is required.

10.2.1.3 Lesson 3: MapReduce is Not Always the Best Choice for Big Data

At the beginning of Dicode, we focused on batch processing. In the final months of the projects, we studied how we can keep the advantages of batch-style distributed systems like Apache Hadoop when dealing with near real-time requirements.

Besides being an obstacle for efficient software development and debugging, the latency of batch processing also interferes with the requirement for "freshness", which is at the core of many information processing applications: in social media monitoring, the user demands an instant alert if there is a new report about the brand or event in question; in a news search, each article has to be analysed and annotated with automatically generated meta-data before being indexed for search. In both cases, a delay of several minutes is not acceptable.

In the literature, a combination of batch and stream processing frameworks is suggested, which combines instant stream processing of incoming data and in-depth processing of batches of data for final results [1]. In Dicode, we evaluated techniques for low latency document analysis relatively late in the project. The envisioned architecture combines batch and stream processing. Large numbers of documents are processed in batch style on Apache Hadoop. Additionally, a fast lane processes high priority documents immediately. This architecture adds additional complexity to the already challenging Hadoop-only solution. Recently, we have seen former users of Hadoop switching completely to stream-processing frameworks. Lightweight stream-processing frameworks like Storm (http://storm-project.net/) now seem to fill a gap in Big Data scenarios and serve as an easier solution for large-scale text mining. In text mining, batch-processing remains important for the re-calculation of statistics, which requires access to the complete data set.

10.2.1.4 Lesson 4: Running a Cluster Consumes Significant Developer Resources

Setup and operations of distributed systems like Hadoop requires profound server and network administration skills, which most software developers in the "Java world" do not have. At the beginning of the project, we underestimated the costs for the administration of the development infrastructure and especially for the skills acquisition in this field. At NEO, we started out with a small cluster of three nodes, because the project did not have any budget for the acquirement of a larger cluster. Later on, we integrated the three machines bought for the Dicode project into a larger cluster shared with another project. By doing so, we could test the technologies developed in Dicode with a more realistic setup. As more people were dependent on the cluster, reliable operations became an issue. Concerning software versions and monitoring tools, both FHG and NEO teams had to develop best practices. Scheduling of jobs and access rights had to be implemented according to the shares of the budget of both projects. Dicode contributed the *Log Aggregation service* (see Chap. 5) for the improvement of operations and debugging. In total, we spent much more developer resources than estimated on the operation of the cluster.

10.2.1.5 Lesson 5: Don't Underestimate the Importance of Meaningful Data Visualization

Text mining is mostly about automatically generating meta-data about documents or document collections. Typically, the resulting annotations are persisted in a data store. In many use cases, text-mining results will not be presented to the end user directly. Typically, the generated meta-data will be used as input for other processing steps. Named Entities, for example, might be indexed by a search engine. The search engine user can then query for Named Entities in a field search.

In other use cases, we want to present the results of our analysis to the end user directly. Most people agree that a good visualisation tells much more than a spreadsheet. But we have to find a good way of presenting the data. Visualization of Big Data can be challenging. Even if the amount of data to be visualized is big, the user should be able to get a general idea about the data on first sight. It should also be easy to switch to a more detailed view, e.g. by zooming into the graph.

Data visualization was not our main focus in Dicode. For us, visualization was mainly a way to present the project results to the partners and to customers of NEO. Our text mining technologies were developed mainly for the integration into back-end components. As all partners agreed on developing the *Dicode Workbench* as a collaboration support platform, we integrated our widgets into it (see Chap. 7). The integration was easy due to the lightweight approach that allowed for the integration of iFrames. In the first year of the project, NEO produced a couple of widgets for the visualization of the Twitter analysis. We used three different chart types: a pie diagram, a zoomable map and a tag cloud. A pie diagram is a typical diagram of the "small data" age. Maps are becoming more and more popular due to the increasing availability of geodata. Tag clouds can be seen as a typical Web 2.0 visualisation—initially they were used to display the ratios of tags assigned by users.

During the project we realized that Big Data visualization had become a very interesting field. Big Data seems to require new types of visualization. People in different disciplines have been experimenting with various new types of diagrams. Today, a tag cloud seems pretty old-fashioned. New types of diagrams emerge: chord diagrams that originally were used in bioinformatics are now used to visualize text mining results (see, for example, http://github.com/norvigaward/naward25/wiki/Babel-2012—Web-Language-Connections).

10.2.1.6 Lesson 6: Twitter's Research Stream is Not Suitable for Social Media Monitoring

Dicode chose Twitter as a major data provider for one of its Use Cases ("Opinion Mining from unstructured Web 2.0 data", see Chap. 9). Many research projects started out analyzing Tweets a couple of years ago. Tweets seemed to be the ideal candidates for Social Media Monitoring. The large amount of accessible Tweets was tempting. In addition, it was obvious that a marketing analyst would have to monitor Tweets because of the near real-time spreading of information, which seemed to be a prototypic case of viral marketing. Twitter also seemed like a good candidate for social network analysis. Twitter's openness towards the developer communities also made the use of Tweets quite attractive.

Today, the situation has changed. Twitter has restricted the API access severely in several ways (http://mashable.com/2012/08/16/twitter-api-big-changes/). Twitter's business model today is mostly based on paid advertisement (so-called "promoted tweets, accounts and trends" (see https://business.twitter.com/marketing-twitter, http://techcrunch.com/2012/01/22/dld-2012-jack-dorsey-twitter-has-a-business-model-that-works/ and http://advertising.twitter.com/)). In addition, Twitter monetizes the data itself – either directly or via companies like Gnip and Datasift.

Our experience in Dicode was that the quality of Twitter's research stream—supposedly 1 % of all global Tweets—is rather low. The daily tag cloud extracted from all Tweets in the research stream mainly shows star signs, computer games or teenager related topics. Pavlo Baron even suspects Twitter of deliberately adding "garbage" to the 1 % research feed [2]. Due to the small sample, applying social network analysis algorithms to the 1 % feed does not make too much sense—for such an analysis, large interconnected amounts of Tweets of a social network of each user would have to be available. A focus on Tweets in German leads to an even sparser dataset.

Various publications deal with the difficulties in mining Tweets. Tweets are short and tend to be idiomatic. This turned out to be a problem for the text mining algorithms implemented in Dicode.

Named Entity Recognition and Disambiguation (NERDist): Our algorithm disambiguates the spotted entity candidates based on the context in which the candidate occurs. In Tweets, there is simply not enough context; our experiments showed that only a small share of Tweets contained more than one entity candidate. In addition, the text in Tweets is not well-formed regarding standard language, which makes it hard to re-use models trained on other text data.

Emotion Detection: The applied Conditional Random Field algorithm extracts positive or negative phrases from documents. Typically, those phrases are as long as a Tweet.

In both cases above, we decided not to adopt the algorithm and/or training to short idiomatic texts. Our reasons are the following: most text mining use cases of NEO and FHG deal with medium sized documents like news. Since both partners want to be able to use the algorithms developed in commercial projects, we decided in replacing Twitter by other text corpora like news and blogs—at least for higher level analyses like NERDist and Emotion Detection. Even if Twitter analysis will be required in a project, we could rely on third-party offers: Twitter data resellers like DataSift already provide sentiment analysis and named entity recognition (http://dev.datasift.com/blog/salience5).

10.2.2 Services

10.2.2.1 Lesson 7: Development Should Be Based on Use Cases

Since the beginning of the project, we worked closely with the use case partners, in order to identify their needs and then convert them into design specifications. The selection of use cases was intended to cover the full range of the features and functionalities of the project, while representing specific domain problems and dealing with various types of large scale and real-time data from heterogeneous sources.

The use case based development of the Dicode services occurred in several steps, each of them was performed in collaboration and interaction with representatives of the use cases. First, the current work practices were reviewed and discussed. The technical partners elaborated the service ideas addressing the problems identified by users. This step enabled us to fully understand the user requirements and discover common characteristics of the use cases regarding the users, the activities and the data. In the second step, prototypes of the single services were created and showed to the use case partners in order to obtain their feedback and to improve the services. The early user feedback helped developers to manage the consequences of design change. In the third step of the service development, several actions were performed concerning the improvement of the services' usability, acceptability and overall quality.

10.2.2.2 Lesson 8: Efforts Put into Data Conversion Tasks Should Not Be Underestimated

Typical data mining software requires an availability of structured data that provide information in a meaningful and descriptive way. The simplest example of the structured data is a table, where the data is stored in columns; one column for each specific attribute and the data is also stored in the row. However, some applications, particularly in specific fields such as Rheumatoid Arthritis Treatment (Dicode's Use Case titled "Trial of Clinical Treatment Effects") prescribe their own requirements to the data format, namely DICOM files (http://en.wikipedia. org/wiki/DICOM). DICOM (Digital Imaging and Communications in Medicine) is a standard for handling, storing, printing, and transmitting information in medical imaging. DICOM files can be exchanged between two entities that are capable of receiving image and patient data in DICOM format.

The original approach for supporting the decision making in the field of Rheumatoid Arthritis Treatment was to apply data mining techniques to doctor's reports about treated patients. Such report is an outcome of the specific software broadly used for analysis of patient data in medical research. This software requires the DICOM files as input format. The analysis of patient data is a very complex process consisting of many steps. In each step, the user is required to interact with the system: to give his feedback concerning classification of patient numbers, to mark region of interest in the image and so on. This analysis process assumes domain knowledge and cannot be reproduced automatically. Therefore, producing a meaningful data set needed for data mining is associated with high manual processing costs. These constitute a bottleneck regarding a user interaction, which was underestimated at the beginning of the project.

Although a large pool of DICOM files was available, the conversion of these files into structured data was not possible because of lack of user sources. The costs for conversion of original data into appropriate formats required by standard tools should not be underestimated.

10.2.2.3 Lesson 9: Real-World Application Needs Analysis of Unstructured Data

The typical Big Data tools assume the availability of structured data, while in many real word applications only unstructured data are available (e.g. texts). This has been the case with the brand watch and marketing applications of a Dicode's Use Case ("Opinion Mining from unstructured Web 2.0 data", see Chap. 9). On the other hand, text-mining algorithms often need to be trained on a set of texts before the "model" can be applied to new texts. Training runs of the algorithms used in the project were notoriously slow with Hadoop when the project started. We therefore decided to integrate sequential training routines in the workbench that operate on text collections of moderate size. However, application of trained models is fast.

10.2.2.4 Lesson 10: Knowledge Extraction Yields Results Which are Often Hard to Interpret

The goal of the Dicode *Subgroup Discovery service* (see Chap. 5) is to generate a human understandable representation of the most interesting dependencies in the data in order to support decision-making in the *Dicode Workbench*. Therefore, the understanding and interpretation of the results are very important issues for the usage of this service. While the user is typically interested in a small yet meaningful output, the outcome of the existing techniques is a huge number of redundant patterns. Having shown the results of the Subgroup Discovery service to researchers, we realised that instead of being supported in knowledge extraction, the user is overwhelmed by the amount of information. Pattern interpretation is a time consuming task, since human experts must manually review the patterns.

In order to reduce the number of raw patterns to a subset of manageable size, we investigated two approaches: one is based on using statistical characteristics; the second includes user feedback in knowledge discovery process. The first approach (described in [3]) uses the statistical quality of pattern to output the k top-quality patterns. This modification leads to a considerable reduction in the amount of returned patterns without losing statistical descriptiveness and, as a consequence, a better understanding of the discovered dependencies. The second approach enables a user to influence the output by including/excluding certain attributes from the search. After reviewing the results, the user can set up a new iteration of the service run by specifying undesired (e.g. biological or medical) attributes. The combination of both approaches enables us to retrieve patterns that have a statistically better quality and, at the same time, are more relevant regarding to the user preferences.

10.2.2.5 Lesson 11: Use Open Source Software Whenever Feasible

Some of the Dicode services are based on existing open source software such as R (http://cran.r-project.org/) and RapidMiner (http://rapid-i.com/). For instance, the *Subgroup Discovery service* uses R to build a connection to Gene Ontology (GO) and to enrich a set of gene names by their functional interpretation, which is given by GO terms.

At the beginning of the Dicode project, we analysed a variety of data mining tools in order to select an appropriate platform for the Dicode data mining services. The outcome of the tool comparison was that R and RapidMiner are freely available open source frameworks that deliver reliable solution for Dicode issues. RapidMiner is the most popular one (even more popular than any commercial product) according to a poll at KDNuggets.com (http://www.kdnuggets.com/polls/2010/data-mining-analytics-tools.html)—a well known and broadly trusted website amongst data miners. R satisfies the most requirements prescribed by the field of biomedical research (which is related to a Dicode's Use Case—see Chap. 8). The experts analysing genomic data have built a wide range of custom libraries for R. Bioconductor (http://www.bioconductor.org/) uses the R statistical programming language and is one of the most popular open source and open development software for the analysis and comprehension of high-throughput genomic data. No one of the existing commercial frameworks offers such flexibility, as it is available with R.

10.2.2.6 Lesson 12: Parallelisation of Data Mining Algorithms May Be Difficult and in Many Situations Even Unfeasible

Some data mining algorithms, e.g. the Subgroup Discovery algorithm, which Dicode *Subgroup Discovery service* is based on, cannot be efficiently parallelized using standard techniques and algorithms. The main challenge associated with parallelization is to break a data mining problem into independent trivial tasks. This requirement cannot be satisfied for the subgroup discovery algorithm. One reliable solution for such kind of problems is in-memory processing. This approach enables an efficient parallelization on the thread level. For better performance, the implemented Subgroup Discovery algorithm exploits the complex in-memory database based on the special data structure called FP-Tree [4]. An

FP-tree is a compact data structure that represents the data set in tree form. Such data representation enables one to reduce both running time and memory size requirements of an algorithm.

10.2.3 User Involvement

10.2.3.1 Lesson 13: User Involvement/Interaction is a Bottleneck

Availability of labelled data is an important assumption for many data mining tasks associated with supervised or semi-supervised learning. In many practical applications, unlabeled instances are abundantly available, while obtaining labelled data is a very costly step, in particular when instance pairs have to be manually labelled by a user. However, incorporation of user feedback in different stages of data mining process enables one to improve significantly the quality of results. For instance, obtaining of relevance feedback is a common practice in information retrieval. The idea of relevance feedback is to involve the user in the retrieval process so as to improve the final result set. In particular, the user gives feedback on the relevance of documents in an initial set of results [5].

The Dicode services which intend to support decision making, such as *Similarity Learning service* (see Chap. 5), require a certain amount of user interaction in order to deliver good results. Our research concerning the problem of how to perform distance metric learning accurately at minimal labeling costs is a reliable solution to avoid the 'user interactions' bottleneck. In Dicode, we proposed a sampling method for selecting a fixed number of 'interesting' instance pairs to label that enabled us to learn a good distance metric [6].

10.2.3.2 Lesson 14: The Organization of Workshops Greatly Helps in Collecting User Feedback

The descriptive nature of local patterns makes them useful as a source of information for decision making. Therefore, the understanding and interestingness of the patterns that are retrieved by the Dicode services are the key requirements for their successful usage. Workshops provide a good opportunity to bring together researchers and practitioners from biology, medicine and bioinformatics domains in order to identify gaps between research and practice and to clarify the user needs in an interactive way. Key questions investigated in a related workshop organized in the context of Dicode were: an appropriate pattern language and inclusion of user feedback in order to improve the interpretation of patterns. Additionally, in the context of the workshop, we organized a challenge where we show the patterns discovered by several algorithms to real experts. We used a questionnaire to obtain the feedback. A detailed analysis of the expert's feedback enabled us to clarify the user needs regarding the interpretation, novelty and interestingness of the discovered patterns. The outcome of the analysis can be summarized as follows:

- The way the results are presented was very important to the user. Very large pattern descriptions are hard to understand.
- Very general patterns are not likely to be useful. Generality of the discovered patterns is not always a characteristic of the data, but include domain knowledge, so the possibility to interactively include the user feedback into discovering process, e.g. to remove very general attributes, is a very helpful function. Moreover, the existing mining algorithms have to be optimized to discover more specific patterns
- Expert knowledge plays an important role—even the best mining methods have to be optimized including the expert feedback.

The expert's feedback provided a significant contribution to the quality and usability of the Dicode services.

10.2.3.3 Lesson 15: Visualisation is Important

Topic models based on the LDA (Latent Dirichlet Allocation) algorithm [7] have been around for several years, but only their visualization in a graph structure was able to bridge the gap between the data-mining expert and the user. In a Dicode's Use Case ("Opinion Mining from unstructured Web 2.0 data", see Chap. 9), the brand watch application greatly benefitted from the visual text collection overview provided in a topic graph.

10.3 Conclusions

There are quite a number of lessons learned in the context of Dicode. The reported lessons have a different scope and importance. Some are quite project specific, while others go far beyond. Future projects may benefit by taking these into account, as is the case for the project partners concerning their individual research and working procedures.

Clearly, the focus of the Dicode project concerning data mining in data-intensive setting has been on batch processing and on improving batch processing using Hadoop. As reported in this chapter, batch processing should be complemented by real-time analysis. This particularly holds for use cases such as public opinion monitoring where a batch cycle of a day or even half a day may be too slow to react immediately, for instance to curtail a "shit-storm". More applications in this area and a deeper understanding of the necessary architecture need to be developed, complemented by an analysis of how to map common data mining tasks and algorithms for this architecture.

References

1. Marz, N., Warren, J.: Big Data—Principles and Best Practices of Scalable Real-Time Data Systems. Manning Publications, New York (2012)
2. Baron, P.: Big Data für IT-Entscheider. Riesige Datenmengen und moderne Technologien gewinnbringend nutzen, München (2013)
3. Grosskreutz, H., Paurat D.: Fast and memory-efficient discovery of the top-k relevant subgroups in a reduced candidate space. In: Machine Learning and Knowledge Discovery in Databases. Lecture Notes in Computer Science vol. 6911, pp. 533–548. Springer, Heidelberg (2011)
4. Han, J., Pei, J., Yin, Y.: Mining frequent patterns without candidate generation. In: Proceedings of SIGMOD'00. pp. 1–12. ACM Press, New York http://doi.acm.org/10.1145/342009.335372 (2000)
5. Büttcher, S., Clarke, C., Cormack, G.: Information Retrieval: Implementing and Evaluating Search Engines. MIT Press, Cambridge, Mass (2010)
6. Friesen, N., Rüping, S.: Distance Metric Learning for Recommender Systems in Complex Domains. In: Proceedings of dicoSyn 2012 (Mastering Data-Intensive Collaboration through the Synergy of Human and Machine Reasoning), a workshop at CSCW 2012, February 12, 2012, Seattle (2012)
7. Blei, D.M., Ng, A.Y., Jordan, M.I.: Latent dirichlet allocation. J. Mach. Learn. Res. **3**, 993–1022 (2003)

Chapter 11
Collaboration and Decision Making in Data-Intensive and Cognitively-Complex Settings: Lessons Learned from the Dicode Project

Spyros Christodoulou, Manolis Tzagarakis, Nikos Karacapilidis,
Fan Yang-Turner, Lydia Lau and Vania Dimitrova

Abstract This chapter reports on practical lessons learned during the development of innovative collaboration and decision making support services in the context of the Dicode project. These lessons concern: (i) the methodology followed and process carried out for the development of the abovementioned Dicode services (ii) the facilitation and enhancement of collaboration and decision making in data intensive and/or cognitively complex settings, and (iii) related technological and integration issues. Detailed evaluation reports, interviews and discussions within the development teams, as well as analysis of the use of the developed services by end-users through the associated log files, provided valuable feedback for the formulation and compilation of these lessons. By sharing insights gained in the context of the Dicode project, this chapter aims to help people engaged in developing similar services.

S. Christodoulou (✉) · M. Tzagarakis · N. Karacapilidis
University of Patras and Computer Technology Institute & Press "Diophantus",
26504 Rio Patras, Greece
e-mail: shristod@cti.gr

M. Tzagarakis
e-mail: tzagara@upatras.gr

N. Karacapilidis
e-mail: nikos@mech.upatras.gr

F. Yang-Turner · L. Lau · V. Dimitrova
University of Leeds, Leeds LS2 9JT, UK
e-mail: F.Yang-Turner@leeds.ac.uk

L. Lau
e-mail: L.M.S.Lau@leeds.ac.uk

V. Dimitrova
e-mail: V.G.Dimitrova@leeds.ac.uk

N. Karacapilidis (ed.), *Mastering Data-Intensive Collaboration and Decision Making*,   213
Studies in Big Data 5, DOI: 10.1007/978-3-319-02612-1_11,
© Springer International Publishing Switzerland 2014

Keywords Collaboration · Decision making · Sense making · Dicode workbench · Dicode services · Visualization · Usability · Software development · Software integration

11.1 Introduction

Dicode's Collaboration and Decision Making Support services (see Chap. 6) constitute innovative solutions that make it easier for users to follow the evolution of an ongoing collaboration, comprehend it in its entirety, and meaningfully aggregate data in order to resolve the issue under consideration in data-intensive situations. These services enable the synchronous and asynchronous collaboration of stakeholders through adaptive workspaces, and serve alternative data visualization schemas. In addition, they facilitate (both individual and group) sense- and decision-making by supporting stakeholders in locating, retrieving and meaningfully interact with relevant information. These services were developed to address the requirements and needs of the project's use cases. To do so, they focused on challenging issues that include the data intensive and cognitively complex characteristics of the use cases, as well as the need for integration with other services (e.g. the Dicode's Data Mining services) in order to further support end users.

In the next section, we report on a series of practical lessons learned while developing and deploying Collaboration and Decision Making Support services in the context of Dicode. Our aim is to share insights and thus assist developers of such services in similar settings. Some of the lessons reported are related to the development of CommBAT, a standalone Windows application ('Community Behavior Analytics Tool'), which supports users to use the log data of Dicode Collaborative Workspace in order to explore and understand collaborative sensemaking behavior (see Chap. 6).

11.2 Lessons Learned

The lessons reported in this chapter are classified according to three perspectives: (i) the process and methodologies followed for developing collaboration and decision making support services (ii) the facilitation and enhancement of collaboration and decision making in data intensive and/or cognitively complex settings, and (iii) the technologies used to implement and integrate the developed services. These lessons can be considered as 'best practices' for people involved in building similar applications. We note that some of the lessons reported below are not exclusively related to the Dicode's Collaboration and Decision Making Support services; they may also be valid for other categories of Dicode services (e.g. the Dicode's Data Mining services—see also Chap. 10).

11.2.1 Software Development Methodology and Process

In this section, we present the outcome of our experiences with regard to the software development methodology and processes. By the term 'software development methodology and processes' we refer to the structuring, planning, and controlling the process of software development in the context of the Dicode project.

11.2.1.1 Lesson 1: Agile Development Methodologies are Well Suited for Teams Developing Independently Software that Needs to Be Integrated into a Single Product

At the beginning of the Dicode project, it was decided to adopt agile development methodologies for managing the Dicode services development. More precisely, the project proposed the use of the Scrum software development framework (see http://www.scrumalliance.org/) that emphasizes cyclical development with rather short feedback loops called 'sprints'. During a sprint, a potentially shippable product increment is created. In general, the duration of such sprints is decided by the development team.

The development team of Dicode's Collaboration and Decision Making Support services fully embraced the Scrum framework. The duration of each sprint was decided to be about 2 weeks. During each sprint, specific functionalities of the abovementioned services were implemented and immediately tested in order to assess their ability to be integrated. Such organization of the development process proved to be important in the context of the project's integration tasks (collaboration and decision support services were designed to be integrated into the Dicode Workbench). The adopted agile methodology was properly supporting the project's integration efforts. Such short sprints were crucial in identifying weaknesses and shortcomings and helped immensely in adjusting and configuring the developed services in the context of the adopted integration architecture. They also helped developers to respond quickly in identified bugs and fine tune the integrated services according to the needs of the project's use cases.

11.2.1.2 Lesson 2: Frequent Meetings with all Technical Partners and Detailed Meeting Minutes Provide the Way to Address Complex Integration Issues

During the Dicode project, all technical partners agreed to conduct regular meetings to discuss and decide on the design and implementation of the foreseen services. In the course of the project, eight technical committee meetings were conducted and detailed minutes were kept (minutes were immediately uploaded to the project's wiki). Meetings were held using videoconferencing tools, where technical partners were discussing their current state of the work, the important design decisions they took since the last meeting, as well as solutions to the project's integration issues.

Such meetings and minutes showed to be very important in properly developing and integrating the collaboration and decision support services. In particular, these frequent meetings gave the opportunity to resolve misconceptions as well as redesign the collaboration and decision support services in order to properly and meaningfully interoperate with the rest Dicode services. Furthermore, they provided a project-wide reference framework which helped the synchronization of the development of the Dicode services. Finally, the meeting minutes functioned as design documents that aided the contextualization of all implementation efforts.

11.2.1.3 Lesson 3: Scenarios Play an Important Role to Elucidate the Requirements Analysis and Provide Tailored Support to Address Big Data Concerns

The aim of the Dicode project was to support the needs of specific Use Cases, which were characterized by high volumes of complex data (see, for instance, the Use Cases presented in Chaps. 8 and 9). The Dicode Use Cases spanned three different fields: biomedical research, medical decision making and online marketing. While all three Use Cases are characterized by high volumes of data, each of them had different needs with respect to collaboration and decision making. To further clarify their requirements early in the project, a set of scenarios were outlined. These scenarios illustrated typical situations in each of these application domains. They proved to be important in addressing the particular needs of each Use Case and in developing the appropriate services and configuration mechanisms required.

In addition, scenarios helped in establishing a common view and vocabulary between technical and non-technical project partners when discussing the foreseen services. They also aided in identifying both the parts of the collaboration and decision making support services that should be generic and the parts that should be configurable, thus providing the tailored support for each Use Case. Finally, they helped developers to properly conceive the different big data concerns in each Use Case. This approach led to the development of the appropriate mechanisms to cope with related issues.

11.2.1.4 Lesson 4: A Standardized Way to Discuss and Document Innovative Services Aids the Establishment of a Common Vocabulary Among Developers and Facilitates Integration Tasks

In the context of the Dicode project, services play an important role in properly supporting the different needs of the project's Use Cases. During the initial project meetings, where technical and non-technical partners were discussing the design, role and use of the foreseen services, the term 'service' was used with many different meanings and in different contexts. This led to misconceptions about what the term refers to, even among the technical partners. In particular, the term 'service' was used in discussions to denote the final software as it would be perceived by end-users, as well as the technical software interfaces that would allow other third-party software to invoke its operations.

The technical partners early on in the project attempted to disambiguate the term 'service' in order to properly set the foundations for the design and implementation of the Dicode services. This led to a project-wide standardization of the meaning, description and documentation of the developed services. According to it, all Dicode services were documented in a consistent way. Such a consistent and coherent description of services also helped immensely the design of the technical integration strategy to be followed.

11.2.1.5 Lesson 5: Multi-Disciplinary Collaboration is Critical in the Design of Innovative Tools

To better understand how a group collaborates and makes decisions, we conducted user studies to understand activities that take place in Dicode's Collaborative Workspace. Logs of activities offer an opportunity to track how users try to make sense, argue and engage others when they try to solve a problem. Our goal was to understand the sensemaking activities of users that lead to a decision and how a collaborative workspace evolves over time.

To reach our goal, we developed CommBAT to support analytics of the log data (see Chap. 6). CommBAT is an analytics tool developed to support collaborative sensemaking research. The development of such a tool requires expertise not only from data analysts, but also from experts in collaborative sensemaking, visualization and semantic technologies. Data analysts contribute to the design on how CommBAT should be used and what aspects of the data are important for the analysis. Researchers in collaborative sensemaking provide theoretical support on the process of collaborative sensemaking. A sensemaking model is useful for the interpretation of the log data. Experts in visualization contribute to the presentation of the data and user interaction techniques. Semantic technologists help to build a rich semantic model to provide a foundation of data analytics. As a working prototype, CommBAT has made a progress towards understanding collaborative sensemaking behaviour with efforts of a team of multi-disciplinary expertise. The next step is to deepen the collaboration and work with wider communities to evaluate its features and improve the design.

11.2.2 Facilitation and Enhancement of Collaboration and Decision Making

This subsection reports on experiences gained with respect to the facilitation and augmentation of collaboration and decision making in data-intensive and/or cognitively complex contexts.

11.2.2.1 Lesson 6: Innovative Metaphors of Collaboration, Although Useful, May Confuse Users and Should Be Introduced in a Way that is Close to What Users are Expecting

Whenever innovative metaphors to collaboration are provided to users, these must be carefully introduced, as there lurks the danger of tool rejection due to encountering of a new and unexplored 'territory'. In general, when users get to use collaborative systems, they expect (based on their experience) to see functionalities offered in wikis, discussion forums and tagging systems, as these are the prevailing applications nowadays on the Web. Our experience showed that radical new ways to collaboration may initially cause confusion rather than excitement.

To address such concerns, the related Dicode services offered the ability to render the collaboration in a way that is familiar to users, by providing alternative collaboration and decision making 'views'. Dicode even enabled a forum-like view of the collaboration (called time-order-view) that displayed the discussion in a way that is found in traditional Web-based discussion forums. This functionality proved to be very helpful as the spatial metaphor of collaboration workspaces (i.e. the "mind-map view") was regarded as simply another way of viewing and conducting an ongoing collaboration (for a detailed description of such functionalities, see Chap. 6).

11.2.2.2 Lesson 7: Alternative Views of Collaboration May Significantly Tame the Complexity of Data-Intensive Workspaces. In Such Environments, Formality in Managing Collaboration Should Not Be Considered as a Predefined and Rigid Property, But Rather as an Adaptable Aspect that Can Be Modified to Meet the Needs at Hand

Generally speaking, existing collaboration support tools provide only a fixed set of abstractions, with which participants may express their opinion and allow only one way of visualizing the associated discourse. More specifically, participants' interaction is regulated by procedures that prescribe and—at the same time—constrain their work. This may refer to both the system-supported actions a user may perform (e.g. types of discourse or collaboration acts), and the system-supported types of collaboration objects (e.g. one has to strictly characterize a collaboration object as an idea or a position). In many cases, users have also to fine-tune, align, amend or even fully change their usual way of collaborating in order to be able to exploit the system's features and functionalities. While such approaches to supporting collaboration are in general useful and used today in a wide range of situations, they are problematic when used in data intensive settings. Traditional collaboration support systems (such as Web-based forums) fail to cope with the great number of items that are uploaded and discussed. One reason for the lack of the proper support is the fixed nature of the available abstractions and static visualization options. There is much evidence that an inflexible set of abstractions often resulted in failures [1].

To overcome such concerns in the Dicode project, the Collaboration Support services were designed to enable multiple views of the same collaboration discourse. Each view introduces a different set of abstractions and a unique way to visualize the collaboration. Dicode's collaboration support services provide the following views (for details, see Chap. 6):

- *Discussion-Forum view*: a collaboration space is displayed as a traditional web-based forum, where posts are displayed in ascending chronological order. Users are able to post new messages to the collaboration space, which will appear at the end of the list of messages. The aim of this view is to allow the collection and sharing of opinions without limiting the expressiveness of participants.
- *Mind-Map view*: a collaboration space is displayed as a 'mind map', where users can interact with the items uploaded so far. The map deploys a spatial metaphor permitting the easy movement and arrangement of items on the collaboration space. The aim of this view is to support 'information triage' [2], i.e. the process of sorting and organizing through numerous relevant materials and organizing them to meet the task at hand.
- *Neighbourhood view*: this view displays a specific item along with its neighbours. The "neighbourhood" of a specific collaboration item is defined as the set of items with which this item is directly connected via a relation in the "Mind-Map view". The aim of this view is to allow users to focus on directly connected items and not be distracted by others.
- *Formal view*: this view enables the posting of predefined knowledge items, which adhere to a specific argumentation model. It invokes a set of dedicated scoring and reasoning mechanisms aiming to aid users conceive the outcome of a particular collaborative session and receive support towards reaching a decision.

Every collaboration workspace in Dicode can be operated and visualized in any of the above views. Users may also switch from one view into another, in order to visualize differently the discourse and use a more appropriate set of functionalities to manage the collaboration items. By doing so, they can also easily spot issues that need attention. Such a flexible way of visualizing and working with the discourse proved to be important in data intensive settings, as the environment could adapt to the increasing number of available resources. It was admitted that alternative views may also reveal previously unseen and potentially valuable insights. Furthermore, these views are based on a unifying conceptual framework which permits an 'incremental formalization' of discourses [3]. Incremental formalization of collaboration proved to be a successful approach in the context of Dicode. In our approach, formality and the level of knowledge structuring was not considered as a predefined and rigid property, but rather as an adaptable aspect that can be modified to meet the needs of the tasks at hand. By the term formality, we refer to the rules enforced by the system, with which all user actions must comply. Allowing formality to vary within the collaboration space, a stepwise and controlled evolution from a mere collection of individual ideas, arguments, annotations and resources to the production of highly contextualized and interrelated knowledge artifacts and actual decisions, can be effortlessly achieved.

11.2.2.3 Lesson 8: Collaboration and Decision Making Services Should Not Be Regarded as 'Application Islands'. Seamless Interoperability is a Crucial Factor for their Adoption and Success

From the users' initial needs and ongoing feedback, openness and seamless interoperability appeared to be a primary need and constitute one of the biggest challenges of today's social software applications [4]. Users want to gain "real ownership" over the information that they have provided and/or belongs to them (e.g. their profile information, projects, and friends). They want to be able to easily import/export data from one environment to another. They want to be able to synchronize information across different tools and visualize it in different ways via different applications. Thus, a seamless integration of distributed tools and services is instrumental when developing innovative collaboration solutions. As a plethora of resources are already available on the Web, collaboration services must explicitly address issues regarding the integration of these resources into their environments. Otherwise, the danger of becoming isolated may surface and ultimately lead to their rejection.

In this direction, the related Dicode services not only facilitate the synchronous and asynchronous collaboration of stakeholders through adaptive workspaces, but they can also efficiently handle the representation and visualization of the outcomes of the data mining services (through alternative and dedicated data visualization schemas) and enable the orchestration of a series of actions for the appropriate handling of data. In addition, they provide an interactive mechanism for indexing and searching of standard documents.

11.2.2.4 Lesson 9: Effective Collaboration and Decision Making Requires Appropriate Mechanisms Tailored to the Needs of Each Use Case

Collaboration and decision making services in Dicode, apart from providing a number of generic mechanisms and functionalities, have been carefully tailored to meet the needs of each one of the three project's use cases. Concerning the collaboration support services, each stakeholder can express his concerns and thoughts through collaboration objects of diverse object types (each collaboration object, apart from its content, carries a specific semantic reflected by its object type). In addition to the predefined collaboration objects types, stakeholders may use a number of different sets of object types dedicated to each specific Use Case.

Regarding the decision making services, the selection of the implemented decision making algorithms was based on a questionnaire filled in by senior decision makers acting in diverse data-intensive settings. According to the results of this questionnaire, decision making algorithms are highly related to the specific problem under consideration. Depending on the specific problem, decision makers require support from algorithms that: (i) allow compensation among the attributes/ criteria used for the evaluation of the alternatives (i.e. a good performance of an alternative concerning one attribute can compensate for a bad performance concerning another attribute) (ii) allow two or more alternatives to be incomparable, and (iii) do not allow compensation among criteria. Three Multi-Criteria Decision Making algorithms, fulfilling the aforementioned prerequisites, were implemented in the context of Dicode (a detailed description appears in Chap. 6): the Weighted Sum Model (WSM), the Analytical Hierarchy Processing (AHP) and the Lexicographic Decision Making rule (LDM).

11.2.2.5 Lesson 10: Analysis of Data-Intensive Collaboration Requires Innovative and Efficient Tools

Data analytics can be ad-hoc and complex. Different analysts can take different approaches for their tasks: statistics, visualization or intuitive observation [5]. For example, for Dicode workspace log data, we conducted analysis using three different tools: Microsoft Excel for cleaning and filtering the data; Mathematica (http://www.wolfram.com/mathematica/) for statistical analysis; Tableau (http://www.tableausoftware.com/) for data visualization. A typical process of this analysis could be:

- Read a Dicode log file (text format) into Excel to conduct basic cleaning and analysis (grouping, sorting or filtering etc.);
- Conduct further analysis in Mathematica, using the output for Tableau or Excel, and
- Read data from Mathematica or Excel into Tableau for visualization.

This process involves a lot of repetitive activities, such as copying and pasting, saving results in different file formats etc. Sometimes, these steps have to be repeated over and over again for different factors considered. To avoid multiple tools and repetitive activities, there is a need for a more efficient tool to support analysts and facilitate the process. To meet this need, we took an innovative approach, developing CommBAT, a tool that enables users to have the features of statistics, visualization and interactivity at one interface. With its features, CommBAT can provide a more efficient way for users to conduct analysis and gain insight of the data. However, it needs to be tested with more researchers and analysts to evaluate its benefit and examine how to bring best practice to the users.

11.2.2.6 Lesson 11: Data Analytics is an Iterative Exploratory Task Which Requires Multi-Perspective View Support

Data analytics is not a 'one-off' but an exploratory activity. The ability to easily extract meaning from complex datasets has become something of a Holy Grail in the tech industry [6]. This is especially true if we intend to gain deep insight about user or group behaviour because it involves lots of different factors. We can look at individual user aspect about who is doing what and in what way; we can look at group aspect on how they interact with each other; we can look at the activity aspect about what activities are involved; we can look at the objects aspect about which object has gained most of the attention. A way to develop insight is to interact with these aspects and look at these factors in different perspectives.

CommBAT has enabled users to interact with all these factors and provide different views of the result. However, there is a lot future work to do to improve the tool, such as free combination of different factors and multiple selections of factor items.

11.2.2.7 Lesson 12: The Need of Rich Semantics Model to Support Design and Analysis of Collaborative Workspaces

One of our important research questions was to understand how a collaborative workspace evolves, for example: how were 'ideas' created? Were they all created at the beginning of the argumentation or at the end of argumentation? To answer such questions, we took a first step to utilize the semantic types to present knowledge objects in 'Object Type view'. Object Type view tried to use the semantic features of Object to visualize the collaborative workspace. However, the current semantic types are quite simple and a rich semantic model is needed to support the design of the tool on how to semantically interpret the data.

11.2.3 Technical and Integration Issues

In this subsection, we refer to technologies used to implement and integrate the developed services. In particular, we focus on our experiences concerning the integration of specific services for collaboration and decision making support, the technology that was used for this integration, as well as the integration of open source frameworks.

11.2.3.1 Lesson 13: Integrating Data Mining into Collaboration Support Services Makes the Collaboration Discourse More Understandable and Greatly Facilitates Collective Sense and Decision Making in Data-Intensive Environments

Integration issues were a central part of the Dicode project. Discussions among project partners were focused on questions such as which integrations are useful in the context of the Dicode use cases, as well as how to technically achieve it. By elaborating the project's use cases and analysing their particular needs with the use case partners, a decision was made to tightly integrate data mining within the collaboration and decision making support services. This was motivated by the observation that in many use cases, such as in the biomedical and marketing domain, data mining and collaboration and decision support services were consistently used in a very specific pattern: in these use cases, stakeholders used collaboration and decision making support services to plan the execution of data mining services as well as to comment on their outcomes. The aim of such integration was to make the implicit relationship of these services explicit, thus facilitating the associated workflow.

This integration allowed data mining algorithms to be first class abstractions in the context of collaboration workspaces—by introducing a new semantic type called "service"- and be part of the discourse elements. This means that data mining algorithms can be used in collaborative discourses as any other discourse element (e.g. 'notes', 'comments' and 'ideas'). Furthermore, data mining algorithms available within a collaboration workspace can be easily executed (even by stakeholders who are not data scientists or analysts) and, after their completion, the results can be automatically uploaded into the collaboration workspace making them available for interpretation and further contemplation. In such a way, stakeholders are able to ask questions of the data based on their own expertise and easily find patterns, spot inconsistencies, or even get answers to questions they have not yet thought to ask.

By allowing the integration of service items into collaboration workspaces and their treatment as any other discourse element, the discourse contextualizes their use, execution and outcomes, thus greatly contributing to its understanding. In addition, as the specific settings under which the services have been executed are stored as part of their metadata (service parameters), the collaboration workspace makes it easy to support the provenance of the results. Such contextualization and understanding were not possible in contemporary data-intensive collaborative environments, which separate the execution of data mining services from their use in the context of collaboration support systems. This ultimately hindered participants from fully comprehending the discourse and made it difficult to assess the origins of the associated resources.

11.2.3.2 Lesson 14: Open Source Visualization Libraries are Mature Enough to Support Visualization in Data Intensive Environments

As the amount of the digital information nowadays is rapidly increasing, one of the major challenges is to make effective use of this vast amount of information. Visual data analysis and information visualization, facilitated by interactive interfaces, enable the detection and validation of expected results, highlight unexpected discoveries in data, validate new theoretical models, provide comparison between models and datasets, enable qualitative and quantitative querying, facilitate decision making and, in general, enable effective data processing and management in data-intensive environments [7].

Driven by: (i) the growing number of open source visualization libraries which have recently emerged (and become popular) (ii) the prerequisites of the Dicode project to embed open source libraries, and (iii) the suggestions expressed in the project reviews, open source visualization libraries have been effectively integrated to provide part of the functionality of the collaboration and decision making support services. In particular, popular open source frameworks have been used to:

- visually outline information and depict the collaboration process in the form of a 'mind-map';
- implement user-friendly components to allow sharing of the collaboration in a number of social software applications;
- provide a graphical representation of the collaboration data and metadata through a number of plots and charts;
- implement a number of supplementary visualization features to enable provenance of information and reduce the data intensiveness in the collaboration environment.

The exploitation of these open source frameworks led us argue that open source visualization libraries are mature enough to support visualization in data intensive environments. Moreover, the required manpower to adapt the related libraries to the needs of the Dicode project proved to be pretty small when compared to the manpower required to implement all functionality from scratch, while the quality of the final implementation remained high.

11.2.3.3 Lesson 15: REST-Based Services Can Support the Tight and Fine-Tuned Integration Required in Data Intensive Environments

From early on, the Dicode project investigated different technologies that can be exploited to implement the required integration of the foreseen services. In particular, we experimented with and compared existing integration architecture styles in the context of the project's data intensive use cases. This review of existing technologies led the technical partners to adopt a REST-based approach to service development and integration.

For instance, the REST-based approach was used for the development of the Dicode workbench (for executing actions such as user logging in a collaboration workspace, creating/updating a collaboration workspace, creating/updating a user's profile, creating a collaboration object from a resource stored in the Dicode repository, etc.). The REST calls implemented for the specific integration proved to be simple and lightweight as they were based on normal HTTP requests. The developer effort required was the minimum possible as no extra toolkits or libraries were involved in the implementation, while the human readable output of the REST calls was a great advantage during the debugging process and the tailoring of the REST calls to meet the exact requirements of the consumer service. The only issue worth mentioning concerns the encryption standards and the algorithms that were used for the data encryption (as the developers of the services used different technologies to implement their services). A close coordination of the developers' efforts was required in order to follow the best possible common standard.

11.3 Conclusions

Developing collaboration and decision making support services to be used in data-intensive and/or cognitively-complex environments is certainly a challenging task. During the design and development of such services in the Dicode project, challenges encountered concerned different areas, such as the development methodology, the facilitation and enhancement of collaboration, as well as the integration technologies. By addressing these challenges, valuable experiences were collected and lessons were learned. For each lesson reported in this chapter, the context in which it arose was presented and its importance was explained.

The overall aim of identifying and discussing these lessons is to share the gained insights with developers who are engaged in developing innovative Web-based collaboration and decision making support services, in order to better structure and plan their work, lessen their development time and deploy such services in a meaningful way in today's data-intensive and cognitively-complex settings.

References

1. Scheuer, O., Loll, F., Pinkwart, N., McLaren, M.B.: Computer-supported argumentation: A review of the state of the art. Int. J. Comput.-Support. Collab. Learn. **5**(1), 43–102 (2010)
2. Marshall, C.C., Shipman, F.M. III: Spatial hypertext and the practice of information triage. Proceedings of the 8th ACM Conference on Hypertext, pp. 124–133, Southampton, 06–11 April 1997)
3. Shipman, F.M., McCall, R.: Supporting knowledge-base evolution with incremental formalization. Proceedings of CHI '94 Conference, pp. 285–291 (1994)
4. Buytaert, D.: From infinite extensibility to infinite interoperability [Online]. http://buytaert.net/from-infinite-extensibility-to-infinite-interoperability (2008). Accessed 22 July 2013
5. Fisher, D., DeLine, R., Czerwinski, M., Drucker, S.: Interactions with big data analytics. Interactions **19**(3), 50–59 (2012)
6. Giurata, P.: User experience design for data discovery and visualization applications—catalyst resources. http://catalystresources.com/saas-blog/data_visualization_user_interface_design_in_the_age_of_big_data/ (2012)
7. Hansen, C.D., Johnson, C.R., Pascucci, V., Silva, C.T.: Visualization for data-intensive science. In: Tansley, S., Hey, T., Tolle, K. (eds.) The Fourth Paradigm: Data-Intensive Science. Microsoft Research, pp. 153–164. (2009)

Printed in the United States
By Bookmasters